W0230530

DEVELOPMENTS IN ANTIBIOTIC TREATMENT OF RESPIRATORY INFECTIONS

NEW PERSPECTIVES IN
CLINICAL MICROBIOLOGY 4

SERIES EDITOR: W. BRUMFITT

DEVELOPMENTS IN ANTIBIOTIC TREATMENT OF RESPIRATORY INFECTIONS

Proceedings of the Round Table Conference on Developments in Antibiotic Treatment of Respiratory Infections in the Hospital and General Practice, held in the Kurhaus, Scheveningen, The Netherlands, June 15–16, 1980

edited by

RALPH VAN FURTH

Department of Infectious Diseases
University Hospital, Leiden
The Netherlands

1981

MARTINUS NIJHOFF PUBLISHERS

THE HAGUE/BOSTON/LONDON

The organization of this round table conference and the publication of the proceedings were made possible by the gratefully acknowledged financial support received from Abbott N.V., The Netherlands

Distributors:

for the United States and Canada

Kluwer Boston, Inc.
190 Old Derby Street
Hingham, MA 02043
USA

for all other countries

Kluwer Academic Publishers Group
Distribution Center
P.O. Box 322
3300 AH Dordrecht
The Netherlands

Library of Congress Cataloging in Publication Data   CIP

Round Table Conference on Developments in Antibiotic Treatment of Respiratory Infections in the
    Hospital and General Practice (1980: Scheveningen, Netherlands)
    Developments in antibiotic treatment of respiratory infections.

    (New perspectives in clinical microbiology; 4) Includes index.
1.   Respiratory organs – Infections – Chemotherapy – Congresses. 2. Antibiotics – Con-
    gresses.
I.   Furth, Ralph van.   II.   Title.   III.   Series.
RC735.A57R68   1980   616.2'00461   81–9533

ISBN-13:978-94-009-8307-6   e-ISBN-13:978-94-009-8305-2   AACR2
DOI: 10.1007/978-94-009-8305-2

CONTENTS

LIST OF CONTRIBUTORS

Cauwenberge van, P.B., Department of Otorhinolaryngology, Academisch Ziekenhuis, De Pintelaan 135, B-9000 Ghent, Belgium

Davies, R.J., Academic Unit of Respiratory Medicine, St. Bartholomew's Hospital, West Smithfield, London EC1A 7BE, England

Dijkman, J.H., Department of Pulmonology, Academisch Ziekenhuis, Rijnsburgerweg 10, 2333 AA Leiden, The Netherlands

Fraschini, F., Institute of Chemotherapy, Faculty of Medicine, University of Milan, Via Vanvitelli 32, 20129 Milan, Italy

Furth van, R., Department of Infectious Diseases, Academisch Ziekenhuis, Rijnsburgerweg 10, 2333 AA Leiden, The Netherlands

Gould, J.C., Central Microbiological Laboratories, Western General Hospital, Grewe Road, Edinburgh EH4 2XU, Scotland

Grob, P.R., Devonshire House, Station Road, Addlestone, Surrey KT15 2AG, United Kingdom

Kalm, O., Department of Otorhinolaryngology, University Hospital, Fack, S-211 85 Lund, Sweden

Kayser, F.H., Institute of Medical Microbiology, University of Zürich, Gloriastrasse 32, CH-8028 Zürich, Switzerland

Kerrebijn, K.F., Department of Pulmonary Diseases, Sophia Kinderziekenhuis, Gordelweg 160, 3038 GE Rotterdam, The Netherlands

Mattie, H., Department of Infectious Diseases, Academisch Ziekenhuis, Rijnsburgerweg 10, 2333 AA Leiden, The Netherlands

Meenhorst, P.L., Department of Infectious Diseases, Academisch Ziekenhuis, Rijnsburgerweg 10, 2333 AA Leiden, The Netherlands

Meer van der, J.W.M., Department of Infectious Diseases, Academisch Ziekenhuis, Rijnsburgerweg 10, 2333 AA Leiden, The Netherlands

Mouton, R.P., Department of Medical Microbiology, Academisch Ziekenhuis, Rijnsburgerweg 10, 2333 AA Leiden, The Netherlands

Simon, C., Abteilung für Allgemeine Paediatrie, Universitäts-Kinderklinik, Schwanenweg 20, 2300 Kiel 1, Germany

Straeten Van der, M., Department of Internal Medicine, Section Chest Diseases, Academisch Ziekenhuis, De Pintelaan 135, B-9000 Ghent, Belgium

Sundberg, L., Department of Otorhinolaryngology, Centrallasarettet, S-371 85 Karlskrona, Sweden

Waaij van der, D., Laboratory for Medical Microbiology, Academisch Ziekenhuis, Oostersingel 59, 9713 EZ Groningen, The Netherlands

1. INTRODUCTION. THE ROLE OF HOST DEFENCE IN RESPIRATORY INFECTIONS

R. VAN FURTH

In healthy individuals there is an equilibrium between the host and the micro-organisms in his environment, including those on the skin and mucous membranes. An infection occurs when the interaction between the host and these micro-organisms is disturbed (Fig. 1). This equilibrium can be disturbed because the host comes into contact with potentially pathogenic micro-organisms to which he has not yet been exposed and against which he does not yet have sufficient resistance, or because the number of potentially pathogenic micro-organisms on the mucous membranes or in the air has increased, or because the

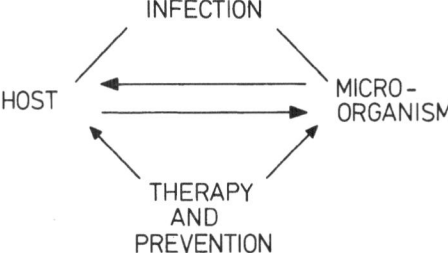

Fig. 1. Interaction between host and micro-organism.

host's resistance has decreased. It must be kept in mind, however, that there is a difference between infection and contamination or colonization. An infection can be defined as a combination of reactions of the host to micro-organisms that have penetrated his body and multiply there. Here, the host plays an active role. Contamination or colonization concerns the presence of living micro-organisms on living tissue or dead material; thus, there is no reaction of the host to the (local) presence of micro-organisms.

MICRO-ORGANISMS AND THE RESPIRATORY TRACT

The respiratory tract is usually divided into the upper respiratory tract, which includes the nose, paranasal sinuses, middle ear, oral cavity, pharyngeal mucous membranes, tonsils, epiglottis, and glottis, and the lower respiratory tract, which comprises the larynx, trachea, in bronchi, bronchioli, and alveoli. This distinction is useful, because under normal conditions the lower respiratory tract below the glottis is sterile, whereas the mouth, nose, and oral cavity are colonized by micro-organisms. If the mucous membranes of the lower respiratory tract have undergone pathological changes (e.g., in chronic bronchitis) or the composition of the mucus is abnormal (e.g., in mucoviscidosis) bacteria can be present on the mucous membranes without inducing inflammatory phenomena. Thus, colonization has occurred but not infection. This means that the presence of only micro-organisms in the sputum is not always an indication of infection, whereas when leukocytes are present as well, a (local) infection is probable if the clinical picture is consistent with this diagnosis.

A wide variety of micro-organisms (viruses, bacteria, fungi, and protozoa) can colonize the upper respiratory tract and are potential pathogens. Micro-organisms which can be involved in respiratory infections are listed in Table 1.

Table 1. Micro-organisms involved in respiratory infections.

Bacteria	Viruses
Streptococcus pneumoniae	Adeno virus
Haemophilus influenzae	
Streptococcus pyogenes	Coxsackie virus
Branhamella catarrhalis	Echo virus
Staphylococcus aureus	Rhino virus
Legionella pneumophila	
Mycoplasma pneumoniae	Corona virus
Corynebacterium diphteriae	
Bordetella pertussis	Influenza virus
	Para-influenza virus
Klebsiella pneumoniae	Respiratory-synocytial virus
Escherichia coli	Measles virus
Proteus species	Mumps virus
Pseudomonas species	
	Cytomegalo virus
Mycobacterium tuberculosis	Epstein-Barr virus
Nocardia asteroides	Herpes simplex virus
	Varicella-zoster virus
Bacteroides species	
Fusobacterium species	Reo virus
Peptostreptococcus species	
	Fungi
Chlamydia psittaci	Candida albicans
Chlamydia trachomatis	Aspergillus fumigatus
	Protozoa
	Pneumocystis carinii

Infections occur when the host is exposed to 'new' micro-organisms against which he has no resistance (antibodies and/or cell-mediated immunity) or is exposed to very large numbers of 'old' micro-organisms for which his defence mechanisms are relatively inadequate, or when the host defence is decreased.

HOST DEFENCE MECHANISMS OF THE RESPIRATORY TRACT

The host possesses a number of mechanisms by which penetrating micro-organisms can be resisted. Usually, air contains about 100 micro-organisms per cubic metre. Inhaled particles larger than 15 μm settle on the nasal mucous membranes and the rear wall of the pharynx, particles smaller than 3 μm reach the alveoli. In first instance the non-immunological defence mechanisms attempt to eliminate the inhaled particles. Examples of disorders of these mechanisms (see Table 2) include abnormal ventilation (e.g., in emphesema, fibrosis), defective cilia function (e.g., in virus infections, Karthagener's syndrome), damaged or altered mucous membranes (e.g., chronic bronchitis, bronchiectasis, influenza and other viral infections, pollution), and abnormal mucus production (e.g. chronic bronchitis, mucoviscidosis). Under such conditions micro-organisms can proliferate locally, penetrate the mucous membrane and underlying tissues (e.g., in pneumococcal pneumonia), and spread, via the blood circulation or lymph stream, to other organs where they remain and multiply (e.g., in pneumococcal sepsis and meningitis).

At the same time, the humoral and cellular immunological defence mechanisms will attempt to inhibit the growth of the micro-organisms and to kill and eliminate them. Humoral factors on the surface of the mucous membrane play an important role in the prevention of infection after colonization. The most

Table 2. Host defence mechanisms in the respiratory tract.

Non-immunological
   Intact mucosal membranes
   Normal ventilation
   Adequate drainage
   Normal secretion and composition of mucus
   Normal ciliary function
   Lysozyme
   Interferon

Immunological
   Normal synthesis of antibodies (secretory IgA; serum IgG and IgM)
   Normal complement system
   Normal numbers and functioning of granulocytes and pulmonary macrophages
   Normal cell-mediated immunity (T lymphocytes and pulmonary macrophages)

important of the humoral factors is secretory IgA (IgA-Sc). This immunoglobulin is produced locally in the mucous membrane and differs in structure from serum IgA, which is single immunoglobulin molecule, by having two IgA molecules with a J chain bound to the secretory component (Sc). Probably due to the presence of the secretory component, secretory IgA is better able to resist degradation by proteolytic enzymes. IgA-Sc does not have the same functions as serum immunoglobulins (IgG, IgA, and IgM), such as a bactericidal action, opsonization of micro-organisms, and complement activation via the classical pathway. IgA-Sc has a number of other important functions, for instance the prevention of adherence of bacteria to the mucous membranes, inhibition of the motility of bacteria, neutralization of viruses and toxins, and activation of the complement system via the alternative pathway. Because of these characteristics, secretory IgA has been called an antiseptic paint, since it forms a layer that protects mucous membranes against the adhesion and penetration of micro-organisms. Some bacteria (Streptococcus sanguis, Neisseria gonorrhoeae, and Neisseria meningitidis) have enzymes (called IgA proteases) which break down IgA_1, and secretory IgA_1. The functions of other immunoglobulins with antibody activity (IgM and IgG) and of the complement factors include the opsonization and killing of bacteria and the neutralization of viruses.

The cellular defence system is formed by granulocytes and alveolar macrophages, which phagocytose opsonized micro-organisms and then kill and digest them. In some forms of infection such as tuberculosis and fungal infections, these phagocytosing cells do not eliminate micro-organisms adequately; only if the alveolar macrophages are stimulated by lymphokines synthesized by sensitized T lymphocytes (i.e., status of cell-mediated immunity) the macrophages can deal effectively with these micro-organisms.

OCCURRENCE OF RESPIRATORY INFECTIONS

It might be expected that as a result of improved socio-economic conditions and the availability of antimicrobial drugs and vaccines, the morbidity and mortality due to respiratory infections would have decreased. There are few statistical data on this point, but the figures in Tables 3 and 4 provide an impression. However, these data represent an underestimation of the actual incidence of respiratory infections because they refer only to hospitalized patients for whom the diagnosis was recorded. The number of patients with respiratory infections who are not admitted and who are seen and/or treated by a general practitioner or as outpatients by a specialist, is much higher. Death due to a respiratory infection must also be higher, because such infections are frequently the cause of death in patients suffering from another disease; in these cases the underlying disease is reported as the cause of death and not the secondary infection.

Table 3. Morbidity due to respiratory infections in The Netherlands*.

	1971		1976	
	Men	Women	Men	Women
Chronic sinusitis	1 941	1 465	2 976	2 312
Pneumonia, bronchopneumonia and influenza	4 154	2 649	6 417	4 067
Bronchitis	5 059	3 153	7 239	4 204
Other diseases of the upper respiratory tract	3 802	2 309	5 926	3 751
Other diseases of the lower respiratory tract	2 606	1 095	4 583	1 910
Total	17 562	10 671	27 141	16 244
Total (corrected)**	27 613	16 778	30 496	18 252

* Absolute numbers; based on data from the Foundation of Hospital Diagnosis Statistics (The Netherlands) 122 hospitals (64%) participated in 1971 and 174 (89%) in 1976; collected by the Central Bureau for Statistics of The Netherlands.
** Linear extrapolation was used to include the other hospitals (100%).

Table 4. Mortality due to respiratory infections in The Netherlands*.

	1971		1976	
	Men	Women	Men	Women
Acute respiratory infections	76	54	62	58
Influenza	136	188	550	636
Pneumonia	1373	1281	1070	1116
Other diseases of the upper respiratory tract	14	6	15	7
Other diseases of the lower respiratory tract	330	198	282	142
Total	1932	1727	1979	1959

* Absolute numbers; based on data from the Central Bureau of Statistics of The Netherlands.

Respiratory infections still occur very frequently, and although treatment with antimicrobial drugs has had a favorable influence on the course of these infections, the mortality is still appreciable. Optimal diagnosis and treatment of these infections are therefore imperative.

THE PRESENT VOLUME

The following are some of the questions raised for the symposium on developments in antibiotic treatment of respiratory infections in the hospital and general practice: What is the present pattern of antibiotic sensitivity to antimicrobial drugs? What is known about the pharmacokinetics of the oral and parenteral antimicrobial drugs in use at present, and what is the degree of penetration of these drugs into lung tissue, mucus, sinus cavities and the middle ear? What is the significance of these levels for the treatment of respiratory infections? Which drugs should be used in the hospital and which in general practice? When is prophylactic treatment appropriate and which drugs should be used for this purpose? What are the side effects of antimicrobial treatment?

This volume contains the introductory talks given at this symposium and the unabridged, largely unedited discussions of the papers.

MICROBIOLOGY

2. THE CURRENT ANTIBIOTIC SENSITIVITY OF HAEMOPHILUS INFLUENZAE

J.C. GOULD

INTRODUCTION

Until recently the sensitivity of Haemophilus influenzae and Pneumococcus to the commonly used antibiotics was regarded as highly predictable. Because of this, and the accepted importance of these organisms as frequent causes of respiratory tract infection, the reliance on bacteriological examinations of secretions of the respiratory tract has been low and the value of laboratory reports regarded as of limited value [1]. However the appearance of strains of H. influenzae more resistant to certain antibiotics, particularly beta-lactamase–producing strains, has focussed attention on variability in antibiotic-sensitivity and the need to carry out more detailed laboratory examinations. These are certainly of great value when the clinician is dealing with blood-borne infection due to Haemophilus spp with or without respiratory infection, for example in meningitis, arthritis or epiglottitis when warning of possible clinical failure can be given and advice on the appropriate antibiotic for treatment.

The importance of H.influenzae and other Haemophilus spp such as H.parainfluenzae, H.haemolyticus H.parahaemolyticus is now more readily accepted [2] and the isolation of these organisms although not difficult requires more sophisticated routine bacteriological methods than are carried out in some diagnostic laboratories.

THE ISOLATION AND CHARACTERISATION OF HAEMOPHILUS ORGANISMS

Sputa and other respiratory tract secretions or swabs sent for bacteriological examination may be examined by direct microscopy. On occasion this may give presumptive evidence of Haemophilus infection if large numbers of small pleomorphic Gram-negative rods are seen in films made form purulent material. Most frequently however bacteriological diagnosis and assessment depends on the demonstration of the organism after growth on suitable culture media and these should be both selective and comparative and enable a quantitative estimate of the number of viable colony forming units present as a measure of the number of organisms present in the specimen.

The best results are obtained by first homogenising sputum by mechanical shaking or by enzymic digestion with pancreatin or trypsin; this allows more even distribution of the organisms throughout the specimen so that those which may occur only amongst the inflammatory material are not missed. Quantitative plating can readily be carried out on the diagnostic plates using a standard loop technique [3] or if more accurate assessment is required, by a suitably modified dilution method after Miles and Misra [4].

Blood-agar, preferably incorporating 10% horse blood and heated-blood agar (chocolate agar), with and without added bacitracin, 10 units per ml [5], are inoculated and the initial area of inoculation ('well area') may be made large enough for this sector of the plate to carry antibiotic-impregnated discs which are useful for differential purposes and for estimating primary antibiotic sensitivity. The results obtained from such primary tests may be more relevant to the individual patient as they are related to the inoculum in this material and may also be affected by other species of bacteria present [6]. Ampicillin-resistance can usually be detected by reduced zones of inhibition around a disc containing 10 μg of the antibiotic although subsequent tests are required to establish that the organism is actually producing beta-lactamase.

H. influenzae will grow with the formation of small colonies after 18–24 hr on the blood agar and can be readily detected within the zone of inhibition of more sensitive flora produced by a penicillin disc containing 1 unit H.parainfluenzae, haemolyticus and parahaemolyticus are seen when present in large numbers by their haemolysis and differentiated from haemolytic streptococci by their limited sensitivity to penicillin and greater sensitivity to amino-sugars, e.g. a disc of gentamycin containing 10 μg. Bacitracin is an excellent selective agent for Haemophilus as at a concentration of 10 units per ml it inhibits most of the other organisms found in the normal respiratory tract thus allowing a more accurate estimate of the number of colonies of Haemophilus spp present.

SUPPLEMENTARY TESTS

A number of tests can be carried out using the primary cultures to establish the 'X' and 'V' dependence of the Haemophilus sp. isolated and whether or not it produces beta-lactamase; the latter may be estimated quickly by the chromogenic cephalosporin test.

Capsular typing, if necessary on strains of H.influenzae may be carried out by slide agglutination using the appropriate Pitman antisera. Sub-culture sensitivity tests are performed using standardised inocula prepared from the primary cultures. Probably the most satisfactory technique with Haemophilus strains is the agar-dilution method using chocolate agar. Known amounts of the antibiotic to be tested are incorporated in freshly prepared medium in a range of concentrations related to tissue levels normally expected during treatment.

Swabs may be treated in a similar manner to sputum specimens but quantitative measurements are more difficult. On arrival at the laboratory the swabs are eluted in sterile buffer solution following which the procedure of bacteriological examination is the same as that given above.

SENSITIVITY OF HAEMOPHILUS TO COMMONLY USED ANTI-BACTERIAL AGENTS

It has become general to express the sensitivity of bacterial pathogens in definite arithmetical terms, i.e. MIC $= \times$ μg per ml. It is however a fact that techniques of estimating the MIC vary from laboratory to laboratory and day to day variation is often wide owing to the large number of variables involved in the tests used, some of which can be readily controlled and some not. It is thus difficult to be sure that results given in this way are comparable and also the presentation of such figures are not always fully meaningful to the clinician for management of the patient. Accordingly it is suggested that more use be made of the index Coefficient of Resistance (CR) [6] which directly compares the sensitivity of the organism under test with a known typical sensitive strain, preferably of the same species. In this way day to day variations in tests are cancelled out and the significance of strains more resistant than normal immediately appreciated by the clinician, that is a CR of greater than 1 indicates more resistance, less than 1 more sensitivity. Clinical usage of this term and correlation with the results of treatment should result in a more accurate definition of what may be considered 'sensitive' and 'resistant' in direct relationship to clinical therapy.

The CR is calculated as follows:

$$\frac{\text{The amount required to inhibit the test organism}}{\text{The amount required to inhibit the control organism}} = \text{CR}$$
to the same degree

At present it is reasonable to suggest that isolates with a CR greater than 10 are 'significantly resistant' and therefore may be refractory to treatment with that agent.

Penicillin. Most strains of Haemophilus are sensitive to between 0.3 and 1.5 μg per ml. With this antibiotic and Haemophilus the comparison should be made with sensitive, clinically susceptible organisms such as penicillin-sensitive Staphylococcus aureus or Streptococcus pyogenes (MIC 0.02–0.06 μg per ml) in which case the CR of all strains of Haemophilus spp is greater than 10. This agrees with much clinical experience that the treatment of Haemophilus influenzae infections with penicillin is unreliable.

Penicillin and Streptomycin. This combination is effective against Haemophilus infections and was successfully used for many years in the treatment of severe respiratory infections. Amino-sugars are generally active against Haemophilus spp.

Ampicillin and Amoxycillin. Until 1978 the great majority of isolates of Haemophilus influenzae were reported as sensitive to ampicillin (MIC 0.3–2.0 µg per ml) with CR less than 10 and this corresponded with good clinical results of therapy with these antibiotics. Since 1968 however, when the first description of a beta-lactamase producing strain was given, the rate of isolation of ampicillin-resistant strains has increased in many centres so that it now stands at between 2 and 10%. Most of these more resistant isolates are resistant because of demonstrable beta-lactamase production and since this is an exocellular enzyme the results of antibiotic-sensitivity tests vary greatly, dependent on the amount of enzyme produced by the organism under the conditions of test. Some infections with these resistant strains have proved refractory to treatment so that the demonstration of beta-lactamase production must be accepted as an indication that the use of ampicillin or amoxycillin may not be clinically effective. Chromosomal-mediated resistance without beta-lactamase production does also occur but the level of resistance may not be high and the CR under test remain less than 10, so that treatment of such strains may be successful.

Tetracyclines. Most strains of Haemophilus spp are sensitive to 0.25–1.25 µg per ml of oxytetracycline and this antibiotic has been effective in the treatment of Haemophilus infections, particularly those of the respiratory tract. A few strains are found to be more resistant with CR greater than 10 but in our experience this proportion has not increased in recent years.

Cotrimoxazole. Trimethoprim is active against Haemophilus spp and its combination with sulphamethoxazole is also active in vitro and in vivo. Only small numbers of more resistant strains are as yet found.

Chloramphenicol. Until very recently all strains of Haemophilus influenzae were found to be sensitive to this antibiotic (MIC 0.5–1.5 µg per ml) and CR less than 10. Thus this agent remains the drug of choice for the treatment of severe infections with blood spread due to Haemophilus spp such as meningitis, endocarditis, arthritis and epiglottitis when it is highly effective, particularly when administered early in the disease. Occasional reports of chloramphenicol-resistance [7, 8] are disturbing and indicate that great care be taken by selected laboratories to monitor isolates for raised CR.

Erythromycin, Lincomycin, Clindamycin and the Cephalosporins. These antibiotics differ from those already mentioned in being less active against Haemophilus spp and each has a larger proportion of strains which are more resistant with CR greater than 10. This has discouraged their routine use for the treament of respiratory infections although many clinical reports indicate that the results of therapy using many of these antibiotics are comparable with therapy using ampicillin, tetracyclines or cotrimoxazole.

Cefotaxime. This is a representative of a new generation of cephalosporins of interest because of its high activity against Haemophilus influenzae and the great majority of strains are sensitive to less than 0.01 μg per ml. On this basis cefotaxime and other new cephalosporins may be highly effective in clinical treatment.

COMMENTARY

Thus Haemophilus influenzae and other Haemophilus are generally sensitive to a wide range of clinically useful antibiotics. If, as seems reasonable, these organisms are important in many patients with upper and lower respiratory tract infection the clinician has a wide choice of drugs that are likely to be effective in treatment and clinical failure is not likely to be frequently associated with in vitro resistance.

It is not yet clear whether excretion of the antibiotic in the secretions of the respiratory tract is related to more efficient elimination of Haemophilus from the respiratory tract and as a result increases clinical efficacy. Most antibiotics are not found in high concentration in sputum and amounts present are marginally related to the in vitro MIC's. This may suggest that tissue concentrations are more important in combating the infecting organisms at the site of inflammation.

If eradication of Haemophilus influenzae is important for rapid clinical recovery and in the prevention of relapse then further study of the effect of combinations of antibacterial agents may be useful. In this context examination for the presence of 'L' forms and their elimination by antibotics may be productive.

Haemophilus spp is important in respiratory infection in the compromised host and the most effective antibiotics may be required for treatment. Thus chloramphenicol, one of the amino-sugars or a new cephalosporin should be considered for such patients.

The increase in resistance due to the spread of plasmid-mediated beta-lactamase production among H.influenzae is a new factor which apart from indicating the need for greater bacteriological surveillance of respiratory infection in the community suggest that an effort should be made to control the use of antibiotics acting as selective agents. This could be the basis of an antibiotic policy for both the general and hospital communities. The large number of clinically active antibiotics now available makes this feasible. Thus erythromycin may be used for a period to the exclusion of ampicillin and other beta-lactam antibiotics so as to preserve the usefulness of the latter and in the hope of reducing the pressure of plasmid selection and spread of resistance throughout the bacterial community. Continuing surveillance of the resistance patterns to the other antibacterial agents must be undertaken by at least a number of reference laboratories using suitable samples of cases of respiratory infection and carrying out sensitivity tests on isolates of H.influenzae and related species.

REFERENCES

1. Percival A: Treatment of respiratory infection. In: Current Antibiotic Therapy, p.151. Churchill Livingstone, 1973.
2. Gould JC: Erythromycin in respiratory tract infections. Scot med J 22:355, 1977.
3. Urquhart GED, Gould JC: Simplified technique for counting the number of bacteria in urine and other fluids. J Clin Path 18:480, 1965.
4. Miles AA, Misra SS: A method of counting bacteria. J Hyg 38:732, 1938.
5. Jarvis JD, Ewins SP: Bacitracin as a selective agent for Haemophilus species. J Med Lab Tech 25:261, 1968.
6. Gould JC, Bowie JH: The determination of bacterial sensitivity to antibiotics. Edin med J 59:178, 1952.
7. Ward JI, Tsai TF, Filici GA, Fraser DW: Prevalence of ampicillin and chloramphenicol resistant strains of H. influenzae. J Infect Dis 138:421, 1978.
8. Peel MM, Tibbits PR, Forsyth JRC: Chloramphenicol-resistant Haemophilus influenzae. Med J Australia 1:130, 1979.

DISCUSSION

Dr Mattie: Dr Gould, I was intrigued by your Coefficient of Resistance (CR) which eliminates a lot of variability in sensitivity testing. But the way you present it, you seem to introduce another kind of variability in comparing different drugs. For instance, how is the sensitive strain defined? Is it so defined that, for different kinds of antibiotics, is has the same meaning? Is that a very sensitive organism, or is it a marginally sensitive organism? I would guess it is very sensitive, because, up to a CR of 10, you can treat patients satisfactorily with antibiotics: that would imply that you would really need 10 times more of the antibiotic in the patient with the marginally sensitive micro-organism, with a coefficient of resistance of 10, than in the very sensitive organism.

So, if at first sight it lookes very attractive, I begin to wonder what the implications are. Especially in comparing different antibiotics and concluding that the degree of resistance is higher for one antibiotic than in the other.

Dr. Gould: Perhaps I did not make it entirely clear. I would normally expect, the control organism to be of the same species. One would take what one regarded as the standard sensitive strain, an organism which has the mean of the more sensitive group. If you have a bi-modal distribution, obviously you would take the more sensitive group.

Originally, when this work was done with Staphylococcus and Streptococcus viridans, the Staphylococcus taken was the standard Oxford Staphylococcus which had an MIC of 0.03 units per ml. Similarly, with the Mycobacterium tuberculosis, it is a standard strain of human mycobacterium that has been used.

Dr Mouton: I would like to comment on this subject, too, because I have the feeling that we are making things needlessly complicated . You say that you do not prefer the CR above MIC, but you used MIC for establishing a certain degree of sensitivity in your control strain. I feel that when you standardized your tests and used your control strain regularly to find out whether your media are alright, the MIC value is much easier to handle than a CR value, which has to be (cor)related for every bacterium separately. So, if your test method is standardized and you use control strains now and then to find out if there is no deviation from the normal values, we would not need this CR value. We would just use MIC as a value which can be compared–for whatever it is worth–with the concentrations that may be achieved in vivo.

Dr Gould: I get your point and I would be the last one to wish to make things more complicated. But I had hoped it had made things somewhat easier, because it is unimportant what the absolute MIC is in this scheme, as long as you have an organism that is reasonably stable. I would have thought that the attraction of the system is that, no matter what the situation with the media or the particular method of testing at any particular time, you have a comparative result. That is, if you have on one occasion an MIC equivalent to 'x' for your control organism and your test organism is '5x'; the next day your control organism happens to be '4x' and your test organism is '10x'. That variation does not matter, since you have still comparative results, the CR, and this enables you to see whether the organism is more or less resistant than your control organism.

Dr Mouton: Don't you feel that the state of the culture itself may influence the results? I mean, the MIC values which you are going to relate with the control strain, and the culture of the control strain itself has an effect on your final results.

Dr. Gould: I would agree with what you say, but in some ways I am avoiding a situation which I think is developing, that is slavish attention given to detailed in vitro tests, with figures for MIC sometimes in small fractions of μg per ml. What does this mean in terms of clinical management? I would go as far as to say, very little. Dr Kayser has already mentioned the statistical significance between one tube difference in an in vitro assessment. I would go further than that and say there is often very little clinical significance in several tubes' difference. I would be the last one to disagree that the differences in CR are of a similar significance. But I think if you have a very large difference, for example, tenfold, then this does have some clinical significance. Perhaps all I'm saying is: let us have a general or crude measurement, nothing too exact in the present state of our knowledge. Maybe a pharmacokineticist will have more to say about this.

Dr Davies: Could I just take that point out from the straightforward clinical point of view. Perhaps we clinicians ought to take more note of what you are doing in the laboratories. For example, Haemophilus influenzae is perhaps the commonest of the respiratory infections that we see in chronic bronchitis. Do you think it is valuable to have sputum cultures and sensitivity tests? Another question I would like to ask is: do you often see primary pneumonia caused by Haemophilus influenzae on its own?

Dr Gould: The answer to your last question is no. We rarely see primary pneumonia due to Haemophilus influenzae. In answer to your first question– which I think was the significance of carrying out bacteriological culture in chronic bronchitis–I think I would subscribe to the procedure of clinical colleagues who treat uncomplicated bronchitis exacerbations empirically. What I would suggest nowadays is that a certain amount of monitoring of these cases in a particular community is helpful in maintaining figures on the sensitivity of Haemophilus influenzae and pneumococci to certain antibiotics, for example, tetracycline for streptococcus pneumoniae and erythromycin and ampicillin for Haemophilus influenzae.

However, there is a further refinement, I think, which is important particularly in the hospital, for patients with chronic bronchitis or similar respiratory syndromes of the lower tract. One can carry out quantitative and properly qualitative bacteriological culture to avoid labeling patients as having gram-negative infections. I am very, very concerned in my own community at the number of chest infections in hospitalized patients that are labeled sometimes as staphylococcal, klebsiella, proteus or pseudomonas infections. These organisms, obviously, have only been upper respiratory-tract colonists. I think that a relatively easy way to overcome this problem is to carry out a semi-quantitative count. If you find that the enterobacteriaciae are present in numbers in the order of 10^6 per ml or less and Haemophilus influenzae are 10^7 to 10^8 per ml or higher, then Haemophilus is clearly more important.

Chairman: Do you wash your sputum before you homogenize it? This is a source of error in determining the flora of the pharynx in unwashed sputum, which is distinct from the flora in the bronchi. You probably know the method of Mulder which consists in washing the sputum with saline and then examining one of the flocs in a gram stained preparation and using another floc for culture.

Dr Gould: Our technique does not include washing. This is not to say I would not approve of that, but, when you are examining a fairly large amount of routine material, this is precluded. All we do is take the sputum as it is sent to us and homogenize it either by ultrasound, or by using digestive enzymes such as trypsin and pancreatin. We have in the past also used shaking and I cannot say I find any great difference.

I think that the question of taking a floc, as opposed to the whole specimen in terms of the actual quantity of organism concerned, is not very important, because even if you are talking about a dilution with the secretions, the nonpurulent material will influence the counts with less than one order of 10, and the techniques one would be following would be unable to detect this.

Dr Mouton: I have a question concerning the type of resistance we find in Haemophilus influenzae now and then. You said you found about 10% Haemophilus resistant to ampicillin. Recently, we have found that most resistant strains of Haemophilus have unusually MIC values above 2, and that they do not produce β-lactamase. Is that your experience, too? In the literature, they are mentioned as a very rare occurrence, but this is not my experience.

Dr Gould: I would agree with you. Our figures are somewhat erratic, partly because of the way in which our resistant strains are obtained. Firstly, I would agree with you that chromosomal-mediated resistance, not dependent on β-lactamase does occur among Haemophilus influenzae strains. I am not sure of its absolute nor of its relative frequency and I would hesitate to quote our figures too widely, because most of our resistance strains at the moment come from a hospital or quasi-hospital environment. For example, a high proportion come from cases of cystic fibrosis that have been treated with antibiotics over a long period of time. As an offshoot of that, we find that there is an increase in frequency of resistance strains in pediatric wards in the hospitals which we are dealing with, which I think is either spread of strains or plasmid transfer. So I am not in a position to say what the true relative or absolute frequency of the chromosomal-mediated resistance is.

Dr Van der Meer: May I comment on that, too? The strains Dr Mouton refers to–the ones that did not produce β-lactamase–were mostly from cystic fibrosis patients and we observed that the sensitivity for carbenicillin was retained. These patients had to be treated therefore with carbenicillin, and with much success. I think this is completely in line with the absense of β-lactamase production.

Is it your experience as well, that you retain carbenicillin sensitivity in those Haemophilus strains?

Dr Gould: In actual fact, the physicians who are treating cystic fibrosis in Edinburgh tend to be concentrated in one unit. They do not use carbenicillin at all. The reason for this, I think, is related to their attitude to the pseudomonas, which very frequently is apparently infecting these children, and the only antibiotic they use, apart from routine ampicillin, is tobramycin. You may have your criticism about that, but I have no control over it.

Dr Kayser: Is trimethoprim allowed to be used alone now, in Great Britain? My second question is: do you know whether in Haemophilus strains, trimethoprim resistance is of the high-resistant plasmid-mediated type?

Dr Gould: The answer to your first question is: Yes. Trimethoprim is now marketed by several companies. I think it is, strictly speaking, only licensed for use in urinary-tract infection in the United Kingdom, but its use is being encouraged quite widely by a large number of persons, microbiologists included. Those who are afraid of the toxic side effects of sulfonamide–and there are a number in my particular region–at the moment are winning the day. Whether or not this will result in a rapid upsurge in trimethoprim resistance, and hence invalidity of cotrimoxazol, remains to be seen.

I am afraid I cannot answer your second question authoritatively. I do not know.

Dr Simon: Did I understand rightly that Haemophilus strains resistant to chloramphenicol were found? How many strains have been found in your laboratory?

Dr Gould: We have seen two strains of Haemophilus influenzae resistant to chloramphenicol. Since 1976, a number of strains–still very, very small in number–resistant to chloramphenicol have been reported. I think this is something that we have to keep in mind.

I have not mentioned the attitude that we have in Edinburgh to blood-borne Haemophilus influenzae infections such as meningitis or arthritis. It is the practice of the Edinburgh clinicians, particularly the pediatricians, to prescribe chloramphenicol until proved wrong. That is, until they have substantiated microbiological evidence to the contrary. This applies to all bacterial meningitides who get triple therapy, including chloramphenicol, until we give them a positive result. They say that this is good because they have good clinical results, not only in Haemophilus infections, but in other meningitides also.

Dr Simon: The two strains you have came from patients with meningitis?

Dr Gould: Yes.

Chairman: Dr Gould, could you as a bacteriologist give an order of preference of antibiotics to be used for respiratory-tract infections. You already alluded to that a little in your lecture, but did not develop a scheme.

Dr Gould: If I may, I think I would still say that for Haemophilus influenzae infections of both the upper and lower respiratory tract, the clinician has a wide selection of drugs including ampicillin or amoxycillin, tetracycline, cotrimoxazol, some of the cephalosporins and erythromycin for treatment. Within my own geographical area, I think it would be left to the clinician to decide.

I think that at least some monitoring should be carried out to make sure that, particularly in the general community, there is no significant rise in resistance. Obviously, if a case is refractory to treatment, then some bacteriological examination should be carried out. If it is true that ampicillin exerts a great deal of pressure in producing plasmids for transfer from resistant to sensitive organisms–and this has already been shown in relation to Staphylococcus by Dr Mouton: that there is a correlation between the pressure that is brought about by the use of one drug in the community and the proportion of resistant organisms–then some sort of policy whereby you can take antibiotics out of use, perhaps in rotation, and preferably then use antibiotics with less in the way of these effects. That is why we have advocated for the general community–at least for a time–the use of erythromycin, which appears to produce less effect in encouraging plasmid transfer of resistance than ampicillin or tetracycline. But, it may be that a drug like erythromycin, if used in great quantity may reveal exactly the same results as the widespread use of ampicillin and tetracycline.

Dr Van der Waaij: It concerns the carriers and the patients frequently relapsing into Haemophilus infections. We have been quite successful with a small number of patients by giving them six tablets of cotrimoxazol per day, suppressing Haemophilus influenzae to a level where it was no longer detectable in the pharyngeal area. In this way, we could induce a rather longer remission period. What is your view regarding the maintenance use of doxycycline, when the patient has either a proven or a clinical suspect upper-respiratory tract viral infection?

Dr Gould: Are you referring to prophylaxis in cases of chronic bronchitis in patients who are liable to exacerbations?

Dr Van der Waaij: Yes.

Dr Gould: I think that the use of tetracyclines for prophylaxis can be successful. Our own results are encouraging, both when it is used over a long period of time, that is, used throughout the whole of the winter which is usually taken as the more likely time for exacerbation and also when used for shorter periods of time, say for a fortnight at a time. It does not work in all patients, but statistically it reduces the number of days off work, which is the usual parameter that we have used among this population. I would think that it has value.

Dr Van der Waaij: Did you see the emergence of resistant bacteria during these long-term treatments?

Dr Gould: Not a great deal. But these were patients who were not in the hospital environment. These were patients who were treated at home and were seen at home. We have followed this routine throughout the years. We do not advocate that this form of treatment be given to persons who are either in hospital or are liable to visit hospital departments.

Dr Van der Waaij: So the use of this drug in low doses is not a selective pressure inside the patient; you feel resistance comes from the environment, from the hospital.

Dr Gould: I think, very largely. This is reflected in a community such as ours by the very low resistance rates in the general population compared with the hospital population.

Dr Sundberg: I would like to ask Dr Gould about the different strains of Haemophilus influenzae. Do you think there is any difference in the properties of capsulated and non-capsulated strains. For example, are strains with a capsule more virulent than strains without a capsule?

Dr Gould: It is an irrefutable fact that virtually every isolate from bloodborne-spread disease due to Haemophilus influenzae or from meningitis is capsulated, and therefore one must ascribe some significance to the capsule. It is also true that in all but perhaps 2 to 4% of isolates from the respiratory tract, whether it be acute exacerbations of chronic bronchitis or an acute pharyngitis, the organisms are non-capsulated. The frequency of capsulation in isolation of Haemophilus influenzae from croup in children is very much higher. I do not think it is 100%, but it is very high. So that I agree there must be some virulence factor associated with the capsule.

However, the evidence is that Haemophilus influenzae and para-influenzae are associated with acute exacerbations of chronic bronchitis and with a number of other respiratory diseases which may well be in every instance secondary to an initial virus, mycoplasma or chlamydia infection. All I am saying is that the organism is there in significant numbers. If we are going to treat the patient, are we going to treat him with an antibiotic for which the organism is susceptible? Do the clinical results match up? The evidence, I think, weighs perhaps to the positive in this case, but it is not irrefutable.

Chairman: To continue along these lines, Dr Gould, you mentioned the L-forms. Do you have any data on the significance of the L-forms. Do they really exist in relapses of chronic bronchitis?

Dr Kayser: I do not believe that L-forms are real pathogens.

Chairman: But they can revert.

Dr Kayser: Ten years ago, there was a lot of talk about L-forms of bacteria, but now no paper has appeared on this subject.

Dr Mouton: I can only say that it is a very difficult question to evaluate, because everybody is convinced that the L-form in itself is not pathogenic, but it can revert and the moment it reverts you are dealing with bacteria which are pathogenic of course. So, what you have to do is to find out the relationship between the isolation of L-forms in patients during remission and the number of exacerbations that occur afterwards.

3. CURRENT PATTERN OF ANTIBIOTIC SENSITIVITY OF PNEUMOCOCCI

F.H. KAYSER

INTRODUCTION

Streptococcus pneumoniae, which can cause respiratory tract infections, meningitis and other types of serious infections, has been known to be exceedingly susceptible to benzylpenicillin and other betalactam antibiotics. Recent reports from South Africa [1, 2], however, of strains resistant to many antibiotics including betalactam drugs, added an alarming new dimension to the antimicrobial susceptibility of Streptococcus pneumoniae. The reports from South Africa indicate that naturally-occurring strains of pneumococci clearly have the potential for significant resistance to penicillin and also to other antimicrobials. About half of the infected patients with bacteremia or meningitis caused by such resistant strains subsequently died. The rapid spread of these pneumococci to other patients and to medical personnel, and their appearance in two hospitals simultaneously, underscore their communicability as well as their virulence.

It is important, therefore, to know the antimicrobial susceptibility of current isolates of Streptococcus pneumoniae elsewhere. However, routine susceptibility testing of clinical isolates has not been considered necessary nor has it been recommended for diagnostic laboratories. Therefore, a number of epidemiological studies about antimicrobial susceptibility patterns of pneumococci in various geographic areas have been performed recently. It is the purpose of this paper to summarize some of the results of these surveys.

RESISTANCE TO ANTIMICROBIALS IN PNEUMOCOCCI

Tetracyclines
It has long been known that resistance of S. pneumoniae to the tetracyclines (MIC of 8 μg/ml or greater) is widespread. Goldstein and co-workers [3] reported an increasing occurrence of tetracycline resistance from 14.3% in 1970 to 39.4% in 1976 in strains isolated in France. Cybulska et al. in Poland [4] found 15% of strains examined to be resistant to the tetracyclines. Recently, Howard et al. [5] reported data obtained in a multicenter study in Great Britain on pneumococcal resistance.

In 1975, 14% of strains were observed to be tetracycline resistant. In 1977, however, the incidence was only 7%. In the United States, tetracycline resistance was found in 4% of the cultures in one study [6] and in 8% in another survey [7]. In Switzerland, 10% of pneumococcal strains isolated in 1978 were found to be resistant to tetracycline [8]. The geometric mean of MICs against resistant strains was 21.4 μg/ml, with a range of MICs between 12.5 and 100 μg/ml. For susceptible cultures, the mean MIC of tetracycline was 0.25 μg/ml [8].

Resistance to tetracycline was observed in more than 50% of the cultures isolated during an epidemic outbreak of pneumococcal disease in two hospitals in Johannesburg, South Africa [2]. These strains also were resistant to penicillin and to other antimicrobials. Since most of the strains belonged to the same serotype, they apparently represented various isolations of the same epidemiologically identical strain. Reference to multiple resistance in pneumococci will be done in a later section of this paper.

Chloramphenicol

Tetracycline was the only antibiotic, against which resistance in significant number of strains were found in various surveys. Resistance to chloramphenicol, for instance, was observed less frequently. Goldstein et al. in France [3] and Cybulska et al. in Poland [4] found resistance to this drug (MIC of 25 μg/ml or greater) in about 5% of their cultures. Howard et al. in 1978 [5] observed only 3 chloramphenicol resistant cultures among 866 strains. In the Switzerland survey [8], one culture out of 180 was resistant to chloramphenicol (MIC: 25 μg/ml). MICs against susceptible strains varied between 0.4 and 12.5 μg/ml with a mean value of 2.9 μg/ml.

Erythromycin and Clindamycin

Resistance to erythromycin (MIC of 8 μg/ml or greater) and clindamycin (MIC of 2 μg/ml or greater) is a rare event in pneumococci. In the epidemiological survey from France [3], only two strains out of 867 pneumococci were found to be resistant to erythromycin. In the studies from the United States [6, 7] and from Great Britain [5], no strains resistant to these drugs were observed.

All pneumococci examined in Switzerland were highly susceptible to erythromycin and clindamycin [8]. MICs of erythromycin were between 0.012 and 0.1 μg/ml (mean: 0.06). Thus, these antibiotics seem to be an excellent alternative to the penicillins, if the latter drugs cannot be used because of resistance or the possibility of allergic reactions in patients.

Cotrimoxazole

Resistance to sulphamethoxazole/trimethoprim is also a rare event in Streptococcus pneumoniae. We observed two strains among our cultures [8], which were inhibited by 25 μg/ml of cotrimoxazole and, therefore, can be considered as weakly resistant. Most of our strains were highly susceptible, the geometric mean of the MICs being 2.4 μg/ml.

Penicillins and Cephalosporins

Recently, resistance to penicillins and cephalosporins has been reported in a strain from the U.S.A. [9] and in a number of cultures isolated in South Africa [1, 2]. Because of concern that these reports might indicate emerging resistance of pneumococci to penicillin on a world wide scale, screening programs to determine whether such strains were prevalent in other geographic areas were undertaken. The results of these newer surveys were in complete agreement with earlier studies [6]. Resistant pneumococci, being inhibited only by 2 μg of benzylpenicillin per ml or more, were not found. [3, 5, 7, 10, 11, 12]. However, there were between 0.1–5.4% of strains with relative decrease in susceptibility to penicillin (MIC 0.1 to 1μg/ml). All strains examined by us in Zürich were susceptible to benzylpenicillin and to cephalothin [8]. The mean MIC of penicillin was 0.01 μg/ml, that of cephalothin 0.16 μg/ml. Penicillin-susceptible pneumococci were also highly susceptible to the ureidopenicillins mezlocillin and piperacillin. The mean MIC of these drugs against a representative number of strains was 0.008 and 0.003 μg/ml, respectively [13].

The newer cephalosporin antibiotics also had excellent results against S. pneumoniae. The geometric mean of MICs of these drugs were as follows: cefamandole: 0.09 μg/ml; cefoxitin: 0.48 μg/ml; cefuroxim: 0.024 μg/ml; cefoperazone: 0.04 μg/ml; cefotaxime: 0.003 μg/ml; moxalactam: 1.19 μg/ml [14].

MULTIPLE ANTIBIOTIC RESISTANCE

Emergence of multiple antibiotic resistance in pneumococci was reported from South Africa in 1977 and 1978 [1, 2] and from the United States in 1978 [9]. The strains identified in Durban 1977 and in Johannesburg 1978, were inhibited by 4–8 μg/ml of benzylpenicillin and by 8–16 μg/ml of cephalothin. In addition strains from Durban were resistant to chloramphenicol (MIC of 9–37 μg/ml) and those from Johannesburg to chloramphenicol (16–32 μg/ml), erythromycin (32–64 μg/ml), clindamycin (64–128 μg/ml), tetracycline (32–64 μg/ml), and cotrimoxazole (20 μg/ml). The strains were susceptible to rifampin, fusidic acid, vancomycin and bacitracin. The strain from the US was resistant to betalactam antibiotics (MIC of penicillin 4 μg/ml and of cephalothin 16 μg/ml) and to tetracycline (MIC 16 μg/ml). Investigations in the mechanism of penicillin resistance showed that the organisms did not produce penicillinase and thus represented cases of intrinsic betalactam resistance [15]. The biochemical basis of intrinsic resistance is not well understood in any bacteria. In a recent study, however, it was shown that penicillin resistance in the South African pneumococci is accompanied by several changes in their penicillin binding proteins, suggesting that resistance to penicillin in these strains involves a number of sequential biochemical alterations [16]. In addititon, it was shown by transformation experiments that high levels of penicillin resistance in these strains cannot be transferred in a single step from resistant donors to susceptible recipients, but only in several consecutive steps, similar to the acquisition of resistance in vitro. From the results it can be concluded that emergence of penicillin resistance in the South African strains may have occurred in the same stepwise selection of mutants, as in test tube experiments, each mutant showing only a small incremental increase in resistance to betalactam antibiotics. The use of low doses of oral betalactam antibiotics for long periods of time may have provided the selective environment for the emergence of these mutants in nature.

CONCLUDING REMARKS

Despite the replacement of serum therapy of pneumococcal disease by the use of potent antimicrobials, infections caused by Streptococcus pneumoniae can still pose a serious problem, and reports from the United States [17] and Great Britain [18] show that the mortality rate in bacteremic pneumococcal pneumonia remains high. The clinical implications, therefore, of pneumococci resistant to antimicrobial agents, particularly to the penicillins and cephalosporins, are serious, and the emergence of penicillin resistant strains in South Africa will modify the approach of clinicians and microbiology laboratories to pneumococcal disease. Recent surveys have shown, however, that multiple resistant strains have not become prevalent elsewhere. At the moment, pneumococci can still be considered to be susceptible to benzylpenicillin and other betalactam antibiotics. Resistance to the tetracyclines occurs in 4–39% of the isolates, resistance to erythromycin, clindamycin, chloramphenicol and cotrimoxazole is noticeable only in occasional strains.

Studies on the mechanism and the genetics of high level penicillin resistance in the South African pneumococci have shown that these strains apparently acquired resistance in a stepwise process, similar to the way in which highly penicillin resistant mutants are obtained in the laboratory. As a possible precaution in holding back the further emergence of resistant strains in nature, the indiscriminate use of betalactam antibiotics, especially in low doses over long periods of time, should be avoided.

REFERENCES

1. Applebaum PC, Bhamjee A, Scragg JN, Hallett AF, Bowen AJ, Cooper RC: Streptococcus pneumoniae resistant to penicillin and chloramphenicol. Lancet ii:995, 1977.
2. Jacobs MR, Koornhof HJ, Robins-Browne RM, Stevenson CM, Vermaak ZA, Freiman I, Miller GB, Whitcomb MA, Isaacson M, Ware JI, Austrian R: Emergence of multiply resistant pneumococci. N Engl J Med 299:735, 1978.
3. Goldstein FW, Dang Van A, Bouanchaud DH, Acar JF: Evolution de la résistance aux antibiotiques des pneumocoques et répartition de leurs types capsulaires. Path Biol 26:173, 1978.
4. Cybulska J, Jeljaszewicsz J, Lund E, Munksgaard A: Prevalence of Diplococcus pneumoniae and their susceptibility to 30 antibiotics. Chemotherapy 15:304, 1970.
5. Howard AJ, Hince CJ, Williams JD: Antibiotic resistance in Streptococcus pneumoniae and Haemophilus influenzae. Report of a study group on bacterial resistance. Brit Med J 1:1657, 1978
6. Cooksey RC, Facklan RR, Thornsbery C: Antimicrobial susceptibility patterns of Streptococcus pneumoniae, Antimicrob Agents Chemother 13:645, 1978.
7. Watanakunakorn C, Glotzbecker C: Susceptibility of recent clinical isolates of Streptococcus pneumoniae to 17 antibiotics. J Antimicrob Chemother 6:83, 1980.
8. Weber F, Kayser FH: Antimikrobielle Resistenz und Serotypen von Streptococcus pneumoniae in der Schweiz. Schweiz med Wschr 109:395, 1979.
9. Cates KL, Gerrard JM, Giebink GS, Lund ME, Bleeker EZ, Lau S, O'Leary MC, Krivit W, and Quie PG: A penicillin-resistant pneumococcus. J Pediatrics 93:624, 1978.
10. Lauer BA, Reller LB: Serotypes and penicillin susceptibility of pneumococci isolated from blood. J Clin Microbiol 11:242, 1980.
11. Maki DG, Helstad AG, Kimball JS: Penicillin susceptibility of Streptococcus pneumoniae in 1978. Am J Clin Pathol 73:177, 1980.
12. Modde HK: Streptococcus pneumoniae isolates relatively insensitive to penicillin G from patients in Switzerland Chemotherapy 24:227, 1978.
13. Kayser FH: Epidemiologie der bakteriellen Resistenz gegen β-Laktam-Antibiotika. In: Berichte über das Internationale Symposium Acylureido Penicilline, Wien 1979. Siegenthaler W, Weuta H, (eds.), Excerpta Medica, 1980, p. 3.
14. Kayser FH: Microbiological studies on cefoperazone. Clin Ther 3, special issue: 24, 1980.
15. Robins-Browne R, Gaspar MN, Ward JI, Koornhof HJ, Jacobs MR, Thornsberry C: Resistance mechanisms of multiple resistant pneumococci: antibiotic degradation studies. Antimicrob Agents Chemother 15:470, 1979.
16. Zighelboim S, Tomasz A: Penicillin binding proteins of multiply antibiotic-resistant South African strains of Streptococcus pneumoniae. Antimicrob Agents Chemother 17:434, 1980.
17. Austrian R, Gold J: Pneumococcal bacteremia with especial reference to bacteremic pneumococcal pneumoniae. Ann. Intern. Med. 60:759, 1964.
18. Calder MA, McHardy VV, Schonell ME: Importance of pneumococcal typing in pneumonia. Lancet i:5, 1970.

DISCUSSION

Dr Gould: Do you have any information about the type of pneumococcus that has been involved in the transformation and, if so, is there any evidence of a change in type concomitant with the increase in resistance?

Dr Kayser: If I refer to the South African pneumococci, these strains were mostly of type 19. The infections caused by these staphylococci were part of an epidemy they had in two hospitals. It was the same strain causing several infections.

Dr Gould: I think I am right in saying that the type distribution in tetracycline resistance is quite wide. There are several types involved, but in South Africa, they are restricted to type 19, are they not?

Dr Kayser: Yes.

Dr Gould: The type 19 remain stable during increase in resistance to benzylpenicillin?

Dr Kayser: Yes, there is no change in serotype.

Chairman: May I ask both of you: do you routinely serotype your isolated pneumococci, or only these rare multi-resistant strains?

Dr Kayser: No, we do not routinely serotype our pneumococci. Lately, we have done this, but not as a service provided by the microbiology laboratory, but as part of an epidemiological study. I do not think it is necessary for the clinician to know what type is causing the disease.

Dr Simon: I remember that South African pneumococci resistant to penicillin-G were highly sensitive to rifampicin. Is this also true of the U.K. and U.S. strains?

Dr Kayser: Yes. As far as this was examined, it is true. It was examined in the United States, but not in the multi-center study from the United Kingdom.

Dr Michel: May I ask Dr Kayser if we could make a little prophecy. I have the impression from the South African reports that the resistant strains which emerged there, arose in an environment making a very indiscriminate use of penicillin. If that is true, then it would mean that in our kind of environment there is not a great chance of such a development of pneumococci resistant to penicillin.

Dr Kayser: Yes, I agree with you completely. It also should be mentioned that, the black population, especially workers in the mines, in which these penicillin-resistant pneumococci in South Africa emerged, is very susceptible (about ten to twenty times more susceptible) to pneumococcal disease, than the white race. This also could have created the environment for these strains to emerge, in addition to the indiscriminate use of penicillin.

Chairman: This higher susceptibility, is it due to the poor socio-economic circumstances and the crowding in the mines? I am not aware of any epidemiological study showing that these groups are different from others.

Dr Kayser: The reason why they are more susceptible is not known. Of course, with the gold miners one reason is the poor socio-economic conditions. We also do not know why pneumococci causes disease; it does not produce any toxins.

Dr Davies: Could I just slightly enlarge on this question from the clinician's point of view? One of the conundrums that I am faced with sometimes is that the pneumococcus appears to be very sensitive to penicillin, and yet all of us experience people with pneumococcal pneumonia who die. You have mentioned the persistent mortality rate that occurs in pneumococcal pneumonia. Certainly people in the United States suggest that there is still a place in some cases for giving antibody. Is this reasonable?

I am actually asking a very general question. I do not understand why some people with pneumococcal pneumonia die. Presumably it has nothing to do with antibiotic resistance.

I also thought there were certain types of pneumococci which were particularly associated with high mortality rates.

Dr Kayser: The reason why, despite the susceptibility to benzyl-penicillin, patients die of pneumococcal pneumonia is because they go to their physician too late. Most of these patients who die come from lower socio-economic groups; they have problems with resistance to infection. Also drug addicts are amongst these patients and they enter the hospital too late.

In a population which can be compared with our normal population in Europe, the mortality rates are much lower. In Switzerland, the mortality rate of pneumococcal pneumonia is about 4%. In the United States, it is reported to be 20 to 30%.

It is true that certain types of pneumococci are causing more often fatal diseases than others. For example, pneumococcus serotype 3 has found a mortality rate of 48%. The reason for this is not known.

Chairman: It might well be that the difference between the United States and Switzerland is medical practice. Pneumococci are the killers in old age, whether the patient goes to the hospital or not.

Dr Dijkman: I could add some recent figures from this country. Pneumococcal pneumonia is comparatively rare here. In our hospitals, we were able to collect about 36 cases in three or four years. Of these, only one died. But about 50% of the patients with pneumococcal pneumonia had some underlying disease.

Dr. Kayser: I once tried to find out the incidence of pneumococcal pneumonia in the community. This is very difficult. I know of only one study in the U.S., which reported that one to two persons per thousand of population per year contract pneumococcal pneumonia. This figure is a little higher than the figure I have suggested for Switzerland, but it is very difficult to get exact figures. We have about 0.5 to 1 pneumococcal pneumonia cases per thousand of population per year.

Dr Mouton: I would like to come back to the problem of the frequency of occurrence of resistance to penicillin. It has been mentioned that it does not occur very often in the Western countries. As far as I know, no resistant strain has been found in Holland, yet. I do not know how it is in Switzerland, but for the time being I feel that even testing for penicillin resistance is not necessary. What is your opinion on this, in view of the rare occurrence?

Dr Kayser: When these reports on the South African pneumococci appeared, we began testing. Up to now, we have not found any resistant strains.

Dr Mouton: Are you continuing this?

Dr Kayser: At the moment, we are still doing it, but I am considering to stop it, because it does not seem necessary to me.

Dr Sundberg: Pneumococci have the ability to persist in tissues and mucous membranes, in spite of an apparently proper antibiotic treatment. Can you explain this phenomenon?

Dr Kayser: No.

Dr Simon: The disc diffusion tests may be erroneous in detecting pneumococci strains which were intermediately sensitive or resistant to penicillin-G. Also, there is a dependence on the kind of medium for testing. What are your recommendations in this connection?

Dr Kayser: We use the agar diffusion disc susceptibility test on Müller-Hinton agar supplemented with 5% blood. For determining penicillin resistance of pneumococci, we use the 5 μg methicillin disc. If the inhibition zone is more than 25 mm around this disc we do an MIC. Up to now, we have not found any resistant or intermediate strains.

Dr Vanderpitte: I would like to comment on your explanation on the higher susceptibility of Africans to pneumococcal infections, I do not agree with you that this is a racial difference. From the literature we know that the children who were victims of this resistant pneumococci were children who had been brought to the hospital for measles. I know from my own African experience that, when a child comes to the hospital with measles, it is a case of malnutrition. So, probably malnutrition is the origin of this higher susceptibility.

A hypothesis has been formulated that multiresistant pneumococci are not as virulent as the normal ones, and that they show their lethal effect just in immunosuppressed or in otherwise predisposed patients. This was examplified by the few sporadic cases of multi- and highly resistant pneumococci isolated in the States; they practically all came from patients with a definitive immunity defect. The hypothesis is that such pneumococci only survive in such patients. So this is a different explanation and it has nothing to do with over-use of penicillin. I do not believe that the South Africans have more access to penicillin than we Belgians have, for instance.

Dr Kayser: I tried to give an explanation of how these strains could have emerged in nature. One fact I mentioned was that pneumococcal infections occur more frequently among this population and it has been speculated that this higher susceptibility of the population is genetically determined. How could these strains have arisen in nature? One possibility is: pneumococci occur more frequently. Another possible explanation: antibiotics are used more frequently and in low doses. For the emergence of resistant strains in nature, low doses have to be used and not high doses. If you use a high dose, you will kill the first- or second-step mutants, but if low doses are used, then the first- or second-step mutants can survive and then a third or fourth mutation can occur and they become progressively more resistant.

Dr Vanderpitte: I agree. There are probably two mechanisms: one bacterial, the step-wise increase of the MICs, and the second to have receptive hosts for them. It is probably not an accident that the second major focus of such pneumococci is in New Guinea, also in people who are very poor and probably in a chronic state of malnutrition.

Dr Mattie: I would like to ask a general question to the speakers of this morning, one which comes up frequently. Is the fact that resistance is the result of beta-lactamase production important, or is it just one of the many causes of resistance? I mean, important from a clinical point of view and for the evaluation of the usefulness of an antibiotic.

Dr Kayser: I think we could organize another symposium to answer this question. It is certain that penicillinase production contributes to resistance, but it is not the only factor. The susceptibility of the target molecules (penicillin-binding proteins) are important also. The access to the target is also an important factor. Penicillinase production, permeability and susceptibility of the target, all taken together, determine the susceptibility or resistance of the whole cell.

The pharmaceutical company that says 'my drug is more resistant to beta-lactamase than your drug, and therefore my drug is better', is wrong.

Dr Gould: I would like to add to what Dr Kayser has said in answer to Dr Mattie. I would agree that the actual mechanism of resistance is immaterial in the clinical management of the patient. But, again referring to the epidemiological approach, I think it has the greatest importance. If the situation is such as has been inferred, with the pneumococcus' step-wise increase in resistance through the action of transformation factors–and this was at one time supposed to be a rather more common mechanism of resistance among micro-organisms–then it should be more easily contained. But it has been obvious with the Gram-negative bacteria that you cannot contain it by simple methods, not even probably by the restriction of one single antibiotic. You have plasmid transfer producing a complete change in the sensitivity of a population within a few hours, particularly in hospitals. I think the appreciation of this factor should have a much greater influence upon the way in which we use and handle antibiotics, particularly in a closed environment.

If I may just add in relation to what has been said about these pneumococci, we did have, I think, one of the first descriptions of a penicillin-resistant pneumococcus from Edinburgh reported in the *British Medical Journal*. There was a certain amount of disbelief at the time about the technology, but it was a true bill. However, in spite of intense observation, at least for a limited period after that, there was no evidence of this isolate being repeated. I wonder if one of the mechanisms whereby this has arisen in the situation described in South Africa is not so much the spread of the resistant strain, as the spread of transformation factors. For example, I would like to know if these children had been examined when they came into the environment. Did they have resistant pneumococci immediately, or did they have sensitive pneumococci which became rapidly resistant on acquiring transformation factors, perhaps successive transformation factors in the hospital. I think this would be very interesting but perhaps very difficult, if not impossible, to find out in retrospect.

Chairman: Thank you for your comment.

Dr Van Boven: When you look to antibiotic sensitivity, it certainly is of great importance in treating patients, especially when bringing this to the level of the individual patient. However, you must realize that 50–60% (to be on the conservative side) of respiratory infection cases occur in general practice, where the treatment is completely empirical. There is no bacteriological investigation. At least in Holland 70% of these cases are treated with antibiotics by the general practitioner.

As said, there is at this moment a whole array of antibiotics available to the general physician. It is very difficult to compare antibiotics purely on the basis of their activity on the bacteria. The sensitivity of the bacteria is, in this respect not relevant, because the groups of patients from whom these data were obtained are not applicable to the general community we are dealing with in this situation.

My question is: how can we initiate studies in order to obtain, in the future, data which is more relevant to the situation, in treating respiratory infections, especially with reference to the general population, so that a general physician could find guidance in the use of antibiotics.

Dr Kayser: I was and still am thinking that the data I presented are relevant for the general physician. These strains were not isolated from hospitalized patients only. All these surveys were done with strains isolated from out-patients as well. In presenting these figures to the physician, he will know what kind of drug he has to use when he suspects a pneumococcal infection. He should not use tetracycline. He can use erythromycin or a beta-lactam antibiotic or cotrimoxazol.

Dr Van Boven: I think it is a very interesting question: in how far the results obtained are data from outpatients that can be seen as representative of the results from the general community. This is the basic question behind this.

Dr Gould: I think it's a very important question. I would agree with Dr Kayser that the results that have been presented this morning, some of them undoubtedly are derived from selected environments, such as hospitals. The results are important in the management of patients in the general community. At the same time I agree that there are difficulties for the general practitioner and he cannot possibly sample all of these cases for bacteriological examination, nor indeed, if he were able to do this, wait for the results before he commences treatment. Practically, one has to compromise and one has to assess the balance. What damage is being done by use of antibiotics empirically when, in terms of increasing resistance, you weigh up the benefits of such empirical therapy? That is, there are fewer cases of bacterial meningitis, there are fewer cases of serious respiratory-tract disease and a number of other infectious conditions, however, haemolytic streptococci of Group A and virulent pneumococci, etc. are in the population in undiminished numbers, compared with the pre-antibiotic era. So that the organisms are there in the community, and I think it is reasonable to suppose that they would cause classical serious disease if it were not for the general use of antibiotics admittedly some of it not specifically indicated in individual patients.

As an answer to the question about what can be done practically: I think that the only way is to monitor properly worked out samples of the population in individual areas, from time to time. This should be the duty of certain reference or regional laboratories. Many do this and they feed to the local medical population the results of their findings. But I think this could be much more generally done.

Dr Mouton: I would like to add a few remarks. I quite agree with the investigation. One should do it in a general population, because we do not know anything about the data on resistance outside the hospitals. I quite agree with Dr Van Boven that the outpatient data are not to be compared with the data in hospitalized patients.

With regard to the three bacteria which have been discussed this morning, I feel that they should not be much of a problem for a general practitioner, because we know which resistances are important to the general practitioner (Staphylococcus aureus to penicillin only, Haemophilus influenzae and pneumococci to tetracycline; even the amipicillin resistance of Haemophilus influenzae does not occur so often in general practice that one has to take it into account, unless of course the patient has been to a clinic and has been treated very often there.)

4. ANTIBIOTIC SENSITIVITY OF STAPHYLOCOCCUS AUREUS. PAST AND PRESENT

R.P. MOUTON

THE EARLY YEARS

The history of resistance of Staphylococcus aureus to antibiotics is almost as old as the use of antibiotics itself: Barber [1], in 1947, described a therapeutic failure in a staphylococcal infection due to penicillin-resistance. Penicillinase production was soon found to be the cause of this resistance [2]. Later it became evident that the gene responsible for penicillinase production is located in a plasmid [3].

There is proof that penicillin resistance existed before the introduction of penicillin [4] and therefore the increase of resistant strains could be totally ascribed to selection of these pre-existing resistant strains. For other antibiotics, selection of chromosomal mutants [5] will often have been the cause. Also the transfer of plasmids with genes coding for a diversity of antibiotic resistance mechanisms, and selection of these variants, on occasions may explain the increased rates of resistance to penicillin and other antibiotics. The transfer of these resistant plasmids by means of transduction indeed appears to be a major cause of multiple resistance [6].

Increases in resistance rates of Staphylococcus aureus for penicillin had already been reported in 1953 [7] and in 1954 an in-patient penicillin resistance rate of 69% was noted in one hospital [8]. Staphylococcal resistance to the tetracyclines also reached levels of up to 52% of the strains at that time. Chloramphenicol resistance did not increase so quickly. In spite of a similar quantitative use of the drug, only 8% of the Staphylococcus aureus strains were resistant [8]. In 1959 Goodier et al. [9] reported penicillin resistance rates of 80% for in-patients, 54% for out-patients, while streptomycin (42%) and sulphonamides (48%) were of decreasing usefulness. Only erythromycin was still at its peak clinical value with 97% sensitivity (Table 1). In The Netherlands, Kooy [10] reported a gradual increase of resistance to four antibiotics (Table 2). Strains derived from the out-patients department of this hospital showed only 11% penicillin resistance. Thus, resistance rates were widely divergent and strongly hospital related.

Multiple resistance also became a common phenomenon. In the report of

Goodier [9] about 45% of the staphylococcal strains was resistant to three antibacterial agents. Hinton and Orr's analysis [8] yielded triple resistance in 11% of the in-patients strains.

In the late fifties it was also observed that the sensitivity patterns of Staphylococcus aureus varied according to the phage type. At that time phage type 81 (group I) strains were mostly resistant to penicillin and tetracycline, while group III strains showed about 50% resistance to these antibiotics. Only a minority of other group I strains and group II strains were resistant [11].

The other antibiotics used for staphylococcal infections were usually active in this period. Resistance to chloramphenicol usually did not increase at the same pace as penicillin, streptomycin or tetracycline, in spite of frequent use of the drug in some clinics [8, 12]. However, increases in resistance rate have been found (Table 3).

For most antibiotics a relationship between use and resistance rate was assumed and this assumption obtained a firmer basis by the experiments with policies aimed at the reduction of prescribing an antibiotic. For instance, a clear cut decrease of resistance rates for all antibiotics was found in a hospital in Germany [13]. Although true for most antibiotics, this relationship between use

Table 1. Antibiotic resistance of 'Staphylococcus aureus in a British hospital'*.

	Year	Month	Percentage of resistant strains to				
			Pen.	Strept.	Tetra.	Chloram.	Ery.
Clinical patients	1956		80	50	30	16	3
	1957	Jan., Febr.	84	54	33	10	5
	1957	Nov., Dec.	81	53	31	10	5
Outpatients	1956		54	18	14	14	2
	1957	Jan., Febr.	64	13	25	6	0
	1957	Nov., Dec.	56	21	18	10	2

* From Goodier and Parry, 1959 [9].

Table 2. Antibiotic resistance of Staphylococcus aureus in an Amsterdam hospital*.

Year	Percentage of resistant strains to				
	Pen.	Strept.	Tetra.	Chloram.	Ery
1954	58	36	17	4	1
1953	66	49	21	7	1

* From Kooy, 1956 [10].

and resistance rate is particularly striking for erythromycin (Table 3). We had a similar experience in 1961 when erythromycin was used to a large extent for the treatment of penicillin resistant staphylococcal infections. With the introduction of methicillin this treatment policy was abandoned and the erythromycin resistance rate of Staphylococcus aureus dropped (Table 4).

BETA-LACTAMASE RESISTANT PENICILLINS

Since the clinical introduction of methicillin in 1961, mention is made of a decline of resistance of Staphylococcus aureus in several reports [14, 15, 16]. Indeed, the almost overall use of methicillin for penicillin-resistant Staphylococcus aureus infections seemed to lead to a spectacular decline of multiple resistance and of erythromycin-resistance [15]. This development was probably aided by the gradually increasing awareness of the need for antibiotic policies.

Perhaps not unexpectedly, the first reports on methicillin-resistance (M.R.) soon appeared [17, 18]. The M.R. strains usually produced large amounts of penicillinase, but this had nothing to do with the M.R. phenomenon, which has been proved to be of an intrinsic nature. The resistance may be chromosomally mediated [19, 20] as well as plasmid mediated [21, 22]. Many studies have been made on the nature of methicillin-resistance [23–27]. Suffice it to conclude from

Table 3. Relationship between resistance rates and the amount of drugs used*.

Year	Month	Erythromycin		Chloramphenicol	
		Resistant strains	Used per month	Resistant strains	Used per month
		(%)	(gram)	(%)	(gram)
1954		10	50	8	300
1955	Jan.–June	16	400	10	800
1955	July–1956 April	11	100	15	1000

* From Hinton and Orr, 1957 [8].

Table 4. Antibiotic resistance of Staphylococcus aureus in a Utrecht hospital.

Year	Percentage of resistant strains to			
	Pen.	Tetra.	Chloram.	Ery.
1959	84	33	12	12
1960	90	36	13	18
1961	81	28	13	13

these studies that in M.R. we are usually dealing with cultures containing several clones with divergent cultural and M.R. properties. In the presence of methicillin the resistant clones consist of a majority of small colony variants and a small percentage of normal colonies. On subculture in the presence of methicillin the latter will be the more common M.R. colony form, with only a small percentage of small colony variants [26]. Recognition of M.R. is usually only possible by culture on 5% NaCl agar plates [28] or by incubation at 30 °C [29].

Right from the beginning the incidence of M.R. Staphylococcus aureus has varied to a great extent. In Denmark [30] and Switzerland [31] M.R. staphylococci became a real problem: in the Zürich area for instance a M.R. rate of 28% was reported for the year 1969 [31]. In London hospitals the M.R. rate was 5% in the same year [29]. Most M.R. strains belong to phage group III and are also resistant to most other antibiotics [31]. In other countries, like the U.S.A., the spread of M.R. strains never occurred to a great extent. The same is true for The Netherlands where, except for some small outbreaks [32, 33], the M.R. rate has been low as a rule. The situation does not seem to have changed much since the late sixties. Outbreaks still occur [34] and there is still some variation in M.R. rates of Staphylococcus aureus in different countries, but the incidence of M.R. infections appears to have fallen in recent years, even in countries with high rates in the sixties [35]. An experience of ours concerning an in-patient in our hospital with continuing M.R. Staphylococcus aureus infections, which did not spread at all in her immediate surroundings (even carriers could not be found) gives support to the suggestion that colonization with these strains may not happen as easily as with other resistant group III strains, at least in the absence of antibiotic treatment.

There are different views regarding the significance of M.R. for treatment. Lacey [36] suggests that the in vitro resistance is not relevant for clinical treatment, but other workers [37, 38] found methicillin to be ineffective in infections with these strains. This has also been our experience. Another important aspect of methicillin resistance is the observation of cross-resistance with the cephalosporins. Unfortunately, the phenomenon is difficult to establish in the laboratory with ordinary techniques. Only prolonged incubation, preferably using large inocula in the tube-dilution technique will make this resistance of the cephalosporins apparent (Table 5) [32]. This cross-resistance has been shown to be important for the treatment of patients [39]. In an experience of our own a patient with a lung abscess caused by M.R. Staphylococcus aureus and treated with cephaloridin, improved for a few days, but then relapsed. At that time a Staphylococcus aureus strain was isolated of which the cephaloridin resistance was much more marked. It is important for microbiologists and clinicians to know that resistances to methicillin and other penicillinase resistant penicillins and to the cephalosporins, run parallel.

Although not particularly important for staphylococcal disease of the lungs,

Staphylococcus epidermidis may be resistant to the penicillinase resistant penicillins [40, 41]. Laverdiere et al. [42] found 56% of these strains to be susceptible to cephalothin, but most of their strains showed high MBC values. However, according to a recent report [43] almost all M.R. Staphylococcus epidermidis strains are also resistant to the cephalosporins, including the new ones like cefamandole, cefoxitin and cefuroxime. The methods used, and in particular the differences in the sized of the inocula, will probably account for these diverging results. Another recent paper [44] stresses the effect of the prophylactic use of penicillinase resistant penicillins and of cephalosporins on the resistance patterns of Staphylococcus epidermidis isolated from the chest before and after operation. Before surgery nearly all strains were sensitive. After operation all strains were resistant to nafcillin and 80% of these were resistant to cefamandole (Table 6). Beta-lactamases do not play any role in this type of resistance.

Table 5. MIC-values of methicillin, cloxacillin and cephaloridin for methicillin-resistant Staphylococcus aureus*.

Inoculum	Methicillin		Cloxacillin		Cephaloridin	
	24 hr**	48 hr	24 hr	48 hr	24 hr	48 hr
	(μg/ml)	(μg/ml)	(μg/ml)	(μg/ml)	(μg/ml)	(μg/ml)
10^5/ml	8	16	0.25	1	0.5	8
10^7/ml	32	>64	1	>16	4	>16

  * From Mouton and Van Boven, 1969 [32].
** Incubation time.

Table 6. Antimicrobial resistance of S. epidermidis isolates from cardiac surgery patients to selected antibiotics*.

Antibiotic		Source of isolate	
		Pre-operative	Post-operative†
	(μg/ml)		
Methicillin	100	0/30**	30/30
Cephalothin	25	0/30	28/30
Cefamandole	25	1/30	24/30
Penicillin G	1.6	4/30	30/30
Streptomycin	100	3/30	20/20
Gentamicin	5	0/30	6/30

  * From Archer and Tenenbaum, 1980 [44].
** Number of isolates resistant/number of isolates tested.
 † 5 days after surgery.

CEPHALOSPORINS

The problems of resistance of Staphylococcus aureus to the cephalosporins are the same as for the penicillinase resistant penicillins. This implies that the resistance rates in Staphylococcus aureus are usually rather low. Then the question about the choice of a cephalosporin remains to be answered. Apart from differences in pharmacokinetics there are also differences in activity of old and new cephalosporins alike. The data on these activities are given in Table 7. We have not been able to find data on systematic susceptibility testing of M.R. Staphylococcus aureus, but we feel that in view of the data in Tables 5 and 7 it may be assumed that the majority of these strains will be resistant to most cephalosporins, just like M.R. Staphylococcus epidermidis.

It is clear from Table 7 that the new cephalosporins do not present any advantage over the older parenteral ones like cephalothin. The oral compounds cephalexin and cephradin are much less active than cephalothin. Cephaloridin appears to be more susceptible to staphylococcal penicillinase than cephalothin [45]. This becomes clear in vitro by using a large inoculum. Against penicillinase negative Staphylococcus aureus cephaloridin is a little more active than cephalothin. However, a serious disadvantage of cephaloridin remains the nephrotoxicity at higher dosage [46].

Table 7. Susceptibility of staphylococci to 11 cephalosporins.

Antibiotic	MIC_{50} values**		
	Staph. aureus	Staph. epidermidis MS*	MR*
	$(\mu g/ml)$	$(\mu g/ml)$	$(\mu g/ml)$
Cephalothin	0.25	0.25	16
Cephaloridin	0.25	0.25	
Cephalexin	8	4	256
Cephradin	8	4	256
Cephazolin	0.25	0.5	32
Cefamandole	0.25	0.5	16
Cefuroxime	1	1	125
Cefoxitin	2	2	125
Cefotaxime	2	4	
Cefoperazone	1	4	
Ly 127935	8	32	

* MS: methicillin sensitive: MR: methicillin resistant.

** Lowest concentrations at which 50% of strains was inhibited. Data collected from several investigations.

TETRACYCLINES

Nowadays the tetracyclines are rarely used for treatment of infections with Staphylococcus aureus. Since many other drugs are available the disadvantages of tetracycline, e.g. bacteriostatic activity and the side effects, particularly regarding the broad spectrum effect on the intestinal flora, preclude their use. Three aspects of tetracyclines and Staphylococcus aureus will be dealt with briefly.

One concerns the fact that tetracycline resistance of Staphylococcus aureus is mostly plasmid linked. Transfer may occur by means of bacteriophages, i.e. by transduction. It is remarkable that, in Staphylococcus aureus, the plasmid concerned only carries the tetracycline resistance-gene [47] and no other genes. Many copies of this plasmid may be present in one cell, the number thereby determining the resistance level. Linkage to other resistance factors is lacking, when the resistance is plasmid mediated. There is evidence that particularly in M.R. Staphylococcus aureus strains, which are mostly tetracycline resistant, the tetracycline resistance determinant is a part of the bacterial chromosome [48]. Then linkage to other resistance genes (erythromycin, streptomycin) may occur.

The second aspect concerns the resistance rates. These tend to vary considerably, but there is a clear tendency to decrease, as is shown in Table 8 [49, 50].

The third feature which should be mentioned is the relatively high activity of minocycline as compared to the other tetracyclines, with regard to Staphylococcus aureus. Minocycline is about 2–4 times as active as doxycycline and about 8 times more active than tetracycline [51]. In spite of this favourable antistaphylo-

Table 8. Percentage of tetracycline resistant strains of Staphylococcus aureus in two localities during the years 1968–1973*.

Year	Zürich		Utrecht	
	Inpatients (%)	Outpatients (%)	Inpatients (%)	Outpatients (%)
1968	27	18	28	12
1969	26	23	23	9
1970			20	9
1971	47	17	15	12
1972	30	10	13	15
1973	38	8	12	9

* From Kayser, 1975 [49] and Mouton et al., 1976 [50].

coccal activity of minocycline, which theoretically might be used in cases of low grade tetracycline resistance, there is complete cross resistance with the other tetracyclines in the sense that strains with high tetracycline MIC-values also show high minocycline MIC-values.

CHLORAMPHENICOL

Some attention has already been paid to the chloramphenicol resistance rate of Staphylococcus aureus during the fifties. We should be aware that comparison of data on resistance which are derived from different sources is prone to errors since techniques and interpretations may differ. This is particularly true for Staphylococcus aureus and chloramphenicol, which is usually marginally active against this species. The reported resistance rates are usually low (< 10%), but variations occur [8, 14, 50, 52], which cannot always be ascribed to differences in use. But there appears to be a relationship between use and resistance rates (Table 3). This has been well illustrated by Phillips [53], who described the occurrence of chloramphenicol resistance of Staphylococcus aureus in one ward which disappeared after retirement of one surgeon who used the drug extensively.

Chloramphenicol resistance in Staphylococcus aureus is usually plasmid linked [54]. This is also the case for Staphylococcus epidermidis, of which species a correlation between chloramphenicol resistance and use has been shown [55]. Like tetracycline resistance plasmids the (small) chloramphenicol resistance plasmid does not carry other resistance genes and may be present in many copies [47]. In spite of the separate occurrence of chloramphenicol resistance plasmids in Staphylococcus aureus, chloramphenicol resistance is usually found in multiple resistant strains.

The mechanism of resistance concerns acetylation by a specific enzyme [56], which is inducible, in contrast to the same type of plasmid determined enzymes in Enterobacteriaceae [47].

MACROLIDES AND LINCOMYCINS

Resistance of Staphylococcus aureus to erythromycin is caused by two different mechanisms. The resistance of strains from clinical sources is usually plasmid mediated and not rarely linked to the penicillinase gene on the same plasmid. In contrast to clinical strains, laboratory strains show combined resistance to erythromycin, oleandomycin and spiramycin [57]. In clinical strains the erythromycin resistance-gene may be linked to other genes than the penicillinase-gene, e.g. the lincomycin- and aminoglycoside-resistance-genes [58]. Erythromycin resistant strains, isolated from patients, easily become resistant to the lincomycins [54]. The MIC of the lincomycins in clinically erythromycin resistant strains is higher than in sensitive strains but when these strains are exposed to lincomycin in vitro the increase of the MIC proceeds at a slower rate than in erythromycin sensitive strains; also M.R. strains tend to acquire resistance to the lincomycins more easily [59].

In line with our own experience several authors [8, 15, 53, 60] found a remarkably rapid decline of erythromycin resistance rates when the use of the antibiotic was restricted, (Table 4). Since the resistance is usually plasmid mediated, this points at a rather unstable plasmid.

Nowadays, erythromycin resistance rates tend to be low (Table 9), in line with low consumption rates. However, when resistance is found we are usually dealing with multiple resistant strains, in which resistance to penicillin, tetracycline, the aminoglycosides and also to the lincomycins (see above) is frequently found.

Epidemiological data on resistance to spiramycin and oleandomycin are lacking, but susceptibility of Staphylococcus aureus to one or both of these macrolides may occur in the presence of erythromycin resistance found in clinical patients [61]. Quite a few strains of this type are found to be resistant to all three drugs. The development of resistance to the macrolides tends to be rather rapid although rarely of importance for the individual patient who is treated with the drug.

There is complete cross-resistance between the lincomycins. The activity of clindamycin for Staphylococcus aureus is about eight times higher than of

Table 9. Resistant strains of Staphylococcus aureus isolated during April 1980 in the University Hospital of Leiden.

	Isolations from	
	Clinical patients* %	Out-patients** %
penicillin	86	88
methicillin	4***	0
erythromycin	2	8
chloramphenicol	2	5
tetracycline	13	10
kanamycin	6	4
gentamicin	0	0

* 104 strains.
** 84 strains.
*** 4 strains from 1 patient.

lincomycin. This implies that low grade resistance to lincomycins may be concurrent with sensitivity to clindamycin.

AMINOGLYCOSIDES

Streptomycin resistance of Staphylococcus aureus is often of the chromosomal type, but plasmid mediated resistance, usually at a lower level (100 μg/ml), occurs. Neomycin and kanamycin resistance may be present in the chromosomally resistant strains, but this resistance is usually plasmid mediated and then streptomycin resistance may be absent. The resistance is due to a phosphotransferase (APH(3′)). Gentamicin resistance is caused by plasmid linked genes; both a phosphotransferase (APH(2″)) and an acetyltransferase (AAC(6′)) have been found to be responsible [62, 63] for modification of the antibiotic. Inactivation appears to be particularly related to the enzyme APH(2″) [64]. A plasmid mediated adenylylating enzyme (ANT(4′)) has also been found which inactivates tobramycin and amikacin, but not gentamicin [65].

Neomycin resistance has often been found in conjunction with frequent topical use of the drug [66]. Crossresistance with kanamycin is important in this respect and has also been confirmed [67] by correlating topical use of neomycin and kanamycin resistance rates. Neomycin resistant staphylococcal infections following the use of topical neomycin have been reported repeatedly [68, 69]. Gentamicin resistance following the topical use of gentamicin has also been described [70]. Frequent parenteral use of gentamicin has led to outbreaks of infections with gentamicin resistant Staphylococcus aureus [34,71].

Kanamycin resistance of Staphylococcus aureus is rare in our hospital, while gentamicin resistance was not found at all in a recent survey (Table 9).

Combined resistance to one or more aminoglycosides with methicillin resistance is rather common [54] and can give serious therapeutical problems [34].

Fortunately, aminoglycoside resistance plasmids appear to be rather unstable [54, 66].

FUSIDIC ACID

Since its introduction in 1962 [72] fusidic acid has been used for multiresistant Staphylococcus aureus infections. Favourable clinical reports on its efficacy in many types of Staphylococcus aureus infection and cystic fibrosis [76] have been made [73, 74, 75].

At an early stage it became apparent that the development of clinical resistance could become a problem since in vitro resistance may be rapidly acquired [72]. Most cultures contain a few primarily resistant variants. Although the risk of Staphylococcus aureus strains becoming resistant in vivo was shown to be smaller than expected [73], resistant strains have emerged during therapy [77]. Combined therapy with other antibiotics has therefore been advocated [75, 78]. In most cases we are dealing with resistant mutants, which occur at the relatively high rate of 10^{-6} and which are selected during therapy. However, primary fusidic acid resistance of Staphylococcus aureus strains has also been found at an early stage [79] and this type of resistance appeared to be linked to other resistances and to be rather unstable, suggesting an extrachromosomal gene. Linkage to the penicillinase plasmid [80] and to tetracycline and kanamycin resistance genes [79] has been described. In particular plasmids with the first type of linkage (PF plasmids) appeared to be spread easily to other strains [81].

Fusidic acid resistance of Staphylococcus aureus is widespread but it is largely dependent on the therapeutic use of fusidic acid, particularly the topical use [82]. Because of the linkage described above, fusidic acid resistance may occur before the drug has been used [83].

There are few studies on resistance rates through the years. Most illuminating are those of Ayliffe et al. [84] for the years 1968–1978, who found negligible resistance percentages in staphylococcal strains from carriers in a general hospital, and 8–22% in dermatology wards, where fusidic acid was used topically. In the burns unit fusidic acid was used sparingly for a few months (2–5% of the patients) in which the fusidic acid resistance rate of Staphylococcus aureus rose from about 5 to 30%; a rapid decrease followed after discontinuation of the drug.

CO-TRIMOXAZOLE

In vitro synergism of sulphamethoxazole and trimethoprim against Staphylococcus aureus strains, which are sensitive to both compounds, can usually be demonstrated [85, 86]; sulphonamide resistant strains rarely show synergy, although trimethoprim may be bactericidal in these cases [86].

Sulphonamide resistance of Staphylococcus aureus will usually be the result of the selection of mutants. Trimethoprim resistance has been shown to be transferable in Enterobacteriaceae [87, 88] as well as in Staphylococcus aureus [89], but in the latter the transfer by means of transduction appears to concern a chromosomal gene [89]. Linkage with sulphonamide resistance has been suggested [90], but this is not a regular phenomenon.

Co-trimoxazole is not routinely used for staphylococcal infections, which probably explains the lack of data on resistance rates. A survey performed in 1977 [91] showed 21.6% of selected multiresistant strains to be resistant to co-trimoxazole. The percentage of resistance in nonselected strains was not determined. Trimethoprim MIC-values varied from 50 to > 100 μg/ml. Sulphonamide resistance of Staphylococcus aureus is much more frequent, but precise recent data are not available. Nakhla in 1972 [92] reported a resistance rate of 18.5% in clinical strains and of 1.6% for trimethoprim in the same collection of strains. M.R. Staphylococcus aureus is always resistant to sulphonamides, but it is usually sensitive to trimethoprim [93].

ANTIBIOTIC TOLERANCE OF STAPHYLOCOCCUS AUREUS

Antibiotic tolerance concerns the phenomenon of normal (low) MIC-values and high bactericidal concentrations of antibiotics that inhibit cell wall synthesis, like the penicillins and the cephalosporins. After the first observation of this characteristic in Staphylococcus aureus [94], several reports on the occurrence of these staphylococcal strains have been made [95, 96]. Watanakunakorn [97] found the susceptibility of tolerant Staphylococcus aureus to be heterogeneous: the majority of cells were killed at the usual concentration but a small number survived. It has been suggested [98] that every culture contains a variable number of tolerant cells, which necessitate antibiotic combination in serious infections.

This characteristic has also been found for antibiotics which do not affect the cell wall like aminoglycosides [97], but 'cross-tolerance' is not always present.

The mechanism of tolerance might be [95] the production of an inhibitor retarding cell autolysis. Recently evidence was found that the property of tolerance may be obtained by lysogenic conversion [99].

The clinical significance of the phenomenon is not clear yet. Slow therapeutical response in infections with tolerant strains has been documented [95], but normal response in serious infections has also been reported [97].

It seems justifiable to ask for a laboratory assessment of tolerance to a therapeutically used antibiotic if there is an unusual delay in clinical response to therapy.

CONCLUSIONS

Antibiotic resistance of Staphylococcus aureus appears to have decreased in the last twenty years. In many instances this favourable development has been obtained by antibiotic policies aimed at decreasing antibiotic usage. Many reports show evident relationships between resistance rates and the use of a particular antibiotic. Varying, but usually low rates of resistance to methicillin and the other beta-lactamase resistant penicillins, cannot always be explained in this way. Multiresistant strains still occur and rarely because of linkage of resistance genes on plasmids. The use of non-related antibiotics to which resistance is not easily evoked in one way or another, appears to be advisable in these cases. Among the newer antibiotics, particularly the new cephalosporins, there is no evidence of superiority over the older ones, with regard to antistaphy-lococcal activity. The clinical significance of antibiotic tolerance remains to be explained.

REFERENCES

1. Barber M: Staphylococcal infection due to penicillin-resistant strains. Brit Med J 2:863, 1947.
2. Barber M: Coagulase positive staphylococci resistant to penicillin. J Pathol Bacteriol 59:373, 1947.
3. Morse ML: Transduction by staphylococcal bacteriophage. Proc Nat Acad Sci USA 45:722, 1959.
4. Smith JMB, Marples MJ: A natural reservoir of penicillin-resistant strains of Staphylococcus aureus. Nature 201:844, 1964.
5. Kayser FM, Wust J, Corrodi P: Transduction and elimination of resistance determinants in methicillin-resistant Staphylococcus aureus. Antimicrob Ag Chemother 2:217, 1972.
6. Novick RP, Bouanchaud D: Extrachromosomal nature of drug resistance in Staphylococcus aureus. Ann N Y Acad Sci 182:279, 1971.
7. Finland M, Haight TH: Antibiotic resistance of pathogenic staphylococci. AMA Archives of Internal Medicine 91:143, 1953.
8. Hinton NA, Orr JH: Studies on the incidence and distribution of antibiotic-resistant staphylococci. J Lab Clin Med 49(2):566, 1957.
9. Goodier TEW, Parry WR: Sensitivity of clinically important bacteria to six common antibacterial substances. Lancet i:356, 1959.
10. Kooy P: Voor antibiotica ongevoelige staphylococcen in een ziekenhuis. Ned T Geneesk 100:2524, 1956.
11. Petersdorf RG, Rose MC: The sensitivity of hemolytic staphylococci to a series of antibiotics. II: A three-year progress report. AMA Archives of Internal Medicine 105:398, 1960.
12. Fisher MW: The susceptibility of staphylococci to chloramphenicol. AMA Archives of Internal Medicine 105:413, 1960.
13. Knörr K, Wallner A: Möglichkeiten zur Bekämpfung des Staphylokokken-Hospitalismus. Dtsch Med Wschr 82:1473, 1957.
14. Bulger RJ, Evans Roberts C, Sherris JC: Changing incidence of antibiotic resistance among Staphylococcus aureus, E. coli, Aerobacter, Klebsiella and Pseudomonas encountered in a teaching hospital over a seven year period. Antimicrob Ag Chemother 1966:42, 1967.
15. Ridley M, Lynn R, Barrie D, Stead KC: Antibiotic-resistant Staphylococcus aureus and hospital antibiotic policies. Lancet i:230, 1970.
16. Ayliffe AJ: Trends in resistance and their significance in primary pathogenic bacteria. In: Recent Advances in Infection (Reeves D and Geddes A, eds.), Churchill Livingstone, Edinburgh, London and New York, 1979, p.1.
17. Jevons MP: 'Celbenin'-resistant staphylococci. Brit Med J I:124, 1961.
18. Cetin ET: Staphylococci resistant to methicillin. Brit Med J II:51, 1962.
19. Sjöström JE, Löfdahl S, Philipson L: Transformation reveals a chromosomal locus of the gene(s) for methicillin resistance in Staphylococcus aureus. J Bacteriol 123:905, 1975.
20. Kayser FM, Wust J, Corrodi P: Transduction and elimination of resistance determinants in methicillin-resistant Staphylococcus aureus. Antimicr Ag Chemother 2:217, 1972.
21. Lacey RW, Grindsted J: Genetic analysis of methicillin-resistant strains of Staphylococcus aureus; evidence for their evolution from a single clone. J Med Microbiol 6:511, 1973.

22. Annear DI, Baron-Hay GS: Loss of resistance to methicillin and other antibiotics in Staphylococcus aureus associated with a chronic empyema. Med J of Australia i:339, 1976.
23. Richmond MH, Parker MT, Jevons MP: High penicillinase production correlated with multiple antibiotic resistance in Staphylococcus aureus. Lancet i:293, 1964.
24. Knox R: Celbenin-resistant staphylococci. Brit Med J I:126, 1961.
25. Barber M: Naturally occurring methicillin-resistant staphylococci. J Gen Microbiol 35:183, 1964.
26. Bulger RJ: In vitro studies on highly resistant small colony variants of staphylococcus aureus resistant to methicillin. J Infect Dis 120(4):491, 1969.
27. Dyke KGH, Jevons MP, Parker MT: Penicillinase production and intrinsic resistance to penicillins in Staphylococcus aureus. Lancet i:835, 1966.
28. Churcher GM: A screening test for the detection of methicillin-resistant staphylococci. J Clin Pathol 21:213, 1968.
29. Parker MT, Hewitt JH: Methicillin resistance in Staphylococcus aureus. Lancet i:800, 1970.
30. Bülow P: Staphylococci in Danish hospitals during the last decade: factors influencing some properties of predominant epidemic strains. Ann N Y Acad Sci 182:21, 1971.
31. Kayser FH, Mak TM: Methicillin-resistant staphylococci. Am J Med Sci 264(3):197, 1972.
32. Mouton RP, Van Boven CP: Tegen methicilline resistente Staphylococcus aureus-infecties in Nederland. Ned T Geneesk 113(3):131, 1969.
33. Michel MF, Priem CC: Control at hospital level of infections by methicillin-resistant staphylococci in children. J Hyg, Camb 69:453, 1971.
34. Crossley K: An outbreak of infections caused by strains of Staphylococcus aureus resistant to methicillin and aminoglycosides. J Infect Dis 139 (3):237, 1979.
35. Rosendal K, Jesson O, Bentzon MW, Bülow P: Antibiotic policy and spread of Staphylococcus aureus in Danish hospitals (1969–1974). Acta pathologica scandinavica Section B, 85:143, 1977.
36. Lacey RW: Can methicillin-resistant strains of Staphylococcus aureus be treated with methicillin? Lancet i:88, 1974.
37. Rountree PM, Beard MA. Hospital strains of Staphylococcus aureus, with particular reference to methicillin resistant strains. Med J Aust 2:1163, 1968.
38. Benner EJ, Kayser FH: Growing clinical significance of methicillin-resistant Staphylococcus aureus. Lancet ii:741, 1968.
39. Chabbert YA: Behavior of 'methicillin heteroresistant' staphylococci to cephaloridine. Postgrad Med J 43 (Suppl.):40, 1967.
40. Sabath LD, Barrett FF, Wilcox C, Gerstein DA, Finland M: Methicillin resistance of Staphylococcus aureus and Staphylococcus epidermidis. Antimicr Ag Chemother 1968:302, 1969.
41. Siebert WT, Moreland N, Williams TW: Methicillin-resistant Staphylococcus epidermidis. South Med J 71 (11):1353, 1978.
42. Laverdière M, Peterson P, Verhoef J, Williams DN, Sabath LD: In vitro activity of cephalosporins against methicillin-resistant, coagulase-negative staphylococci. J Infect Dis 137 (3):245, 1978.
43. John JF Jr., McNeill WF: Activity of cephalosporins against methicillin-susceptible and methicillin-resistant, coagulase-negative staphylococci. Minimal effect of beta-lactamase. Antimicr Ag Chemother 17 (2):179, 1980.
44. Archer GL, Tenenbaum MJ: Antibiotic-resistant Staphylococcus epidermidis in patients undergoing cardiac surgery. Antimicr Ag Chemother 17 (2):269 1980.
45. Benner EJ, Bennett JV, Brodie JL, Kirby WMM: Inactivation of cephalothin and cephaloridine by Staphylococcus aureus. J Bacteriol 90 (6):1599, 1965.
46. Kabins SA, Cohen S: Cephaloridine therapy as related to renal function. Antimicr Ag Chemother 5:922, 1965.
47. Falkow S: The staphylococcal plasmids. In: Infectious Multiple Drug Resistance, Pion Ltd., London, 1975.

48. Kayser FH, Felix M, Wüst J: Extrachromosomal and chromosomal drug resistance in methicillin resistant Staphylococcus aureus. In: Bacterial Plasmids and Antibiotic Resistance, (Krčméry V, Rosival L, Watanabe T, eds.), Spinger-Verlag, New York, 1972, p.371.
49. Kayser FH: Methicillin-resistant staphylococci. Lancet ii:650, 1975.
50. Mouton RP, Glerum JH, van Loenen AC: Relationship between antibiotic consumption and frequency of antibiotic resistance of four pathogens – a seven-year survey. J Antimicr Chemother 2:9, 1976.
51. Steigbigel NH, Reed CW, Finland M: Susceptibility of common pathogenic bacteria to seven tetracycline antibiotics in vitro. Am J Med Sci 255:179, 1968.
52. Krčméry V, Grunt J, Rosival L, Calpaš Š: Antibiotic resistance in various medical services. Infection 7 (2):S216, 1979.
53. Phillips I: Antibiotic policies. In: Recent Advances in Infection (Reeves D, Geddes A, eds.), Churchill Livinstone, Edinburgh London and New York; 1979, p.151.
54. Lacey RW: Antibiotic resistance plasmids of Staphylococcus aureus and their clinical importance. Bact Rev 39:1, 1975.
55. Bentley DW, Hahn JJ, Lepper MH: Transmission of chloramphenicol-resistant Staphylococcus epidermidis: Epidemiologic and laboratory studies. J. Infect Dis 122:365, 1970.
56. Shaw WV, Brodsky RF: Characterization of chloramphenicol acetyltransferase from chloramphenicol resistant Staphylococcus aureus. J Bacteriol 95:28, 1968.
57. Lacey RW: Lack of evidence for mutation to erythromycin resistance in clinical strains of Staphylococcus aureus. J Clin Pathol 30:602, 1977.
58. Grinsted J, Lacey RW: Generic variation of streptomycin-resistance in clinical strains of Staphylococcus aureus. J Med Microbiol 6:351. 1973.
59. Mouton RP, Hos JM: Sensitivity of Staphylococcus aureus to lincomycins in relation to erythromycin and methicillin sensitivity. In: Advances in Antimicrobial and Antineoplastic Chemotherapy, (Hejzlar M, Semonský M, Madák D, eds.), Urban & Schwarzenberg, München Berlin Wien, 1972, p.623.
60. Lilly HA, Lowbury EJL: Antibiotic resistance of Staphylococcus aureus in a burns unit after stopping routine prophylaxis with erythromycin. J Antimicr Chemother 4 (6):545, 1978.
61. Hudson DG, Yoshihara GM, Kirby WMM: Spiramycin: clinical and laboratory studies. AMA Arch Int Med 97:57, 1956.
62. Brown DFJ, Kayser FH, Biber J: Gentamicin resistance in Staphylococcus aureus. Lancet ii:419, 1976.
63. Shannon KP, Phillips I: Gentimicin-resistant Staphylococcus aureus. Lancet ii:580, 1976.
64. Scott DF, Wood DO, Brownell GH, Carter MJ, Best GK: Aminoglycoside modification by gentamicin-resistant isolates of Staphylococcus aureus. Antimicr Agents Chemother 13:641, 1978.
65. Devaud M, Kayser FH, Huber U: Resistance of bacteria to the newer aminoglycoside antibiotics: an epidemiological and enzymatic study. J Antibiotics XXX:655, 1977.
66. Noble WC, Naidoo J: Evolution of antibiotic resistance in Staphylococcus aureus: the role of the skin. Br J Dermatol 98:481, 1978.
67. Starkey DH, Gregory E: Fluctuations in antibiotic resistances of gramnegative rods in one hospital, 1961–70, with comment on kanamycin-neomycin relationship. CMA 105:587, 1971.
68. Lowbury EJL, Babb JR, Brown VI, Collins BJ: Neomycin-resistant Staphylococcus aureus in a burns unit. J Hygiene, Cambr 62:221, 1964.
69. Rountree PM, Beard MA: The spread of neomycin resistant staphylococci in a hospital. Med J Australia i:498, 1965.
70. Wyatt TD, Ferguson WP, Wilson TS, McCormick E: Gentamicin. J Antimicr Chemother 3:213, 1977.
71. Lewis SA, Altemeier WA: Emergence of clinical isolate of Staphylococcus aureus resistant to gentamicin and correlation of resistance with bacteriophage type. J Infect Dis 137:314, 1978.

48

72. Goatfredsen WO, Jahnsen S, Lorck H, Roholt K, Tybring L: Fusidic acid; a new antibiotic. Nature 193:987, 1962a.
73. Crosbie RB: Treatment of staphylococcal infections with 'Fucidin'. Brit med JI:788, 1963.
74. Matsaniotis N, Messaritakis J. Anagnostakis D: Fusidic acid for staphylococcal infections in children. Brit Med JI:564, 1967.
75. Jensen K, Lassen HCA: Fulminating staphylococcal infections treated with Fucidin and penicillin or semisynthetic penicillin. Ann Intern Med 60 (5):790, 1964.
76. Norman AP: Fucidic acid in cystic fibrosis. Lancet ii:216, 1967.
77. Lowbury EJL, Cason JS, Jackson DM, Miller RWS: Fucidin for staphylococcal infection of burns. Lancet ii:478, 1962.
78. Jensen KA, Klaer I: Fucidin. A study on problems of resistance I Acta Pathol Microbiol Scand 60 (2):271, 1964.
79. Evans RJ, Waterworth PM: Naturally-occurring fusidic acid resistance in staphylococci and its linkage to other resistance. J Clin Pathol 19:555, 1966.
80. Chopra I, Bennett P, Lacey RW: A variety of staphylococcal plasmids present as multiple copies. J Gen Microbiol 79:343, 1973.
81. Lacey RW, Rosdahl VT: An unusual 'penicillinase plasmid' in Staphylococcus aureus. Evidence for its transfer under natural conditions. J Med Microbiol 7:1, 1974.
82. Ayliffe GAJ, Green W, Livingston R, Lowbury EJL: Antibiotic resistant Staphylococcus aureus in dermatology and burn wards. J Clin Pathol 30:40, 1977.
83. Naidoo J, Noble WC: Acquisition of antibiotic resistance by Staphylococcus aureus in skin patients. J Clin Pathol 31:1187, 1978.
84. Ayliffe GAJ, Lilly HA, Lowbury EJL: Decline of the hospital staphylococcus? Lancet i:538, 1979.
85. Darrell JH, Garrod LP, Waterworth PM: Trimethoprim: Laboratory and clinical studies. J Clin Pathol 21:202, 1968.
86. Lewis EL, Anderson JD, Lacey RW: A reappraisal of the antibacterial action of cotrimoxazole in vitro. J Clin Pathol 27:87, 1974.
87. Fleming MP, Datta N, Grüneberg RN: Trimethoprim resistance determined by R factors. Brit Med J I:726, 1972.
88. Amyes SGB, Emmerson AM, Smith JT: R-factor mediated trimethoprim resistance: result of two three-month clinical surveys. J Clin Pathol 31:850, 1978.
89. Nakhla LS: Genetic determinants of trimethoprim resistance in a strain of Staphylococcus aureus. J Clin Pathol 26:712, 1973.
90. Lewis EL, Lacey RW: Present significance of resistance to trimethoprim and sulphonamides in coliforms, Staphylococcus aureus, and Streptococcus faecalis. J Clin Pathol 26:175, 1973.
90. Chattopadhyay B: Co-trimoxazole resistant Staphylococcus aureus in hospital Practice. J Antimicr Chemother 3:371, 1977.
92. Nakhla LS: Resistance of Staphylococcus aureus to sulphamethoxazole and trimethoprim. J Clin Pathol 25:708, 1972.
93. Seligman SJ: IN vitro susceptibility of methicillin— resistant Staphylococcus aureus to sulfamethoxazole and trimethoprim. J Infect Dis 128 (suppl):S543, 1973.
94. Best BK, Best NH, Koval AV: Evidence for participation of autolysis in bactericidal action of oxacillin on Staphylococcus aureus. Antimicrob Ag Chemother 6:825, 1974.
95. Sabath LD, Laverdiere M, Wheeler N, Blazevic D, Wilkinson BJ: A new type of penicillin resistance of Staphylococcus aureus. Lancet i:443, 1977.
96. Bradley JJ, Mayhall CH, Dalton HP: Incidence and characteristics of antibiotic-tolerant strains of Staphylococcus aureus. Antimicr Ag Chemother 13:1052, 1978.
97. Watanakunakorn C: Antibiotic-tolerant Staphylococcus aureus. J Antomicr Chemother 4:561, 1978.

98. Bradley HE, Weldy PL, Hodes DS: Tolerance in Staphylococcus aureus. Lancet i:150, 1979.
99. Bradley HE, Wetmur JG, Hodes DS: Tolerance in Staphylococcus aureus: Evidence for Bacteriophage Role. J Infect Dis 141:233, 1980

DISCUSSION

Dr Kayser: Dr Mouton, do you have any idea about the frequency of infections of de lower respiratory tract caused by staphylococci, in relation to other organisms?

Dr Mouton: Of course, the rate is very low indeed. I have no exact data on the number of infections, but I think it must below 5%, perhaps even lower than 1%.

Dr Simon: I am surprised that cefamandole did not act better than cephalothin in methicillin resistant staphylococci. How many strains were investigated?

Dr Mouton: The data were derived from several investigations, including some of our own, so I cannot say exactly how many strains were involved. [Other data on methicillin resistance of Staphylococcus epidermidis are given in Table 6.] We found cefamandole activity against methicillin-sensitive strains, but not against methicillin-resistant strains. Perhaps Dr Kaiser also has this kind of data?

Dr Kayser: Yes. We examined in more than hundred methicillin-resistant strains, the resistance to cephalothin, cefamandole and other antibiotics. The resistance to cephalothin and cefamandole was nearly the same; cefamandole acted a little better, but one dilution in a MIC assay is not significant to me. If I do MIC-determinations with cephalothin and cefamandole and calculate geometric means of the MICs they come out nearly the same.

Dr Mouton: Do you agree with me, Dr Kayser, that one should not use cephalosporins in staphylococcal infections when there is methicillin resistance?

Dr Kayser: I completely agree with you, except if a strain is multiple resistant, i.e., resistant to many other drugs such as tetracyclines, chloramphenicol and erythromycin. Then the clinician may think of using a beta-lactam drug, but then he has to use a good acting beta-lactam drug like cephalothin or cefamandole.

Dr Mouton: So if you have no other choice, you think one can use it?

Dr Kayser: Because of the heterogeneity of the strains, most of the cells causing an infection are killed and only a resistant minority is not. The organism might handle these highly resistant bacteria. In certain cases, it is justified to use a beta-lactam drug.

Dr Mouton: We have seen relapses of staphylococcus lung infections after treatment with cephaloridin, in cases of methicillin resistance. First we saw some beneficial effect, but after two or three days we got a relapse; by then, these strains were completely resistant to cephaloridin too.

Chairman: Dr Kayser, how often do you see a patient like this, with a multiple resistant strain? Especially in respiratory infections, this must be very rare.

Dr Kayser: I cannot give you data about multiple resistant staphylococci in respiratory-tract infections, but as to infections in general I can say that in the University Hospital of Zürich about 5% of the isolated staphylococci exhibit this multiple antibiotic resistance.

Chairman: Which is different from the findings in Leiden.

Dr Kayser: Our figures are much higher than yours are, unfortunately.

Dr Michel: May I ask Professor Mouton if he would give us his definition of a tolerant strain?

Dr Mouton: I do not think I am the person to give that definition. There have been several descriptions of the tolerant strains. I think the ratio of MBC to MIC should be higher than 32. I do not know whether it is quite correct to use absolute values for this characteristic, because it will possibly depend on the way in which a selection has already been made from the total culture.

Dr Michel: The reason why I asked the question is that we are now treating a patient with serious staphylococcus septicemia. The staphylococcus has a ratio of not more than 2 between MBC and MIC. But, if you do a survival curve, it appears that the strain is not killed at all by 8 μg/ml of cloxacillin in 24 h. Would you call that a tolerant strain?

Dr Mouton: I do not think so, but I do not know whether testing bactericidal activity by survival curves is the correct technique. Besides, the activity of a lot of drugs, particularly the penicillins and cephalosporins, should be tested in the first six or eight hours, when using survival curves. After that period you will nearly always find regrowth, even in the case of completely normal MIC and MBC values.

Chairman: Are you satisfied?

Dr Michel: Not completely. But Professor Mouton indicated that we are somewhat uncertain about the whole field of tolerance and that we believe it is an important field for future investigation.

Dr Mouton: I quite agree, but according to the definition that has been given of tolerance, in cases of small difference in MBC and MIC values we are not dealing with tolerance. I do not think that the findings you just described indicate tolerance. As to tolerance itself, its significance is not known.

Dr Van der Meer: Could you formulate under which circumstances the testing in the routine laboratory for tolerant strains or testing for MBC is indicated?

Dr Mouton: When a patient does not react properly to treatment, one should test for tolerance after two or three days of treatment, at least if we are dealing with serious infections. I would not do it for every type of infection. So in case of serious infections, the strains should be kept.

Dr Mattie: It is still not quite certain what the clinical implication of tolerance is.

Dr Mouton: That is true.

Dr Simon: Have you any data about the frequency of resistant strains to cotrimoxazole?

Dr Mouton: No, because we are not testing them routinely. We do not feel that it is necessary for staphylococcus infections to do routine tests for cotrimoxazole. According to the literature the resistance rates are usually rather low. The highest I have been able to find was 20%.

Dr Kayser: May I come back to the tolerance problem? Sometimes treatment of a staphylococcus septicemia with beta-lactam antibiotics is ineffective. We always examine these strains very carefully and some exhibit phenomena of tolerance; then the drug has to be changed and another bactericidal drug has to be used. We also find strains which are completely susceptible, like normal staphylococcus, but despite these in vitro data the clinical outcome is not good. This has something to do with the host; the host is not able to eliminate the bacteria. It is not a question of the antibiotic.

Dr Van der Meer: I have a question for Dr Kayser. What would he give in a situation where he finds no response and no tolerance: A non-responding patient with a severe staphylococcus infection not reacting to beta-lactam antibiotics?

Dr Kayser: There is a possibility to use aminoglycosides and vancomycin. Both antibiotics are bactericidal, but these antibiotics have also negative aspects. Resistance to aminoglyocosides in staphylococci is low, as Dr Mouton has pointed out, and there are no vancomycin-resistant staphylococci as far as I know.

Dr Van der Meer: What about rifampicin? There have been papers about staphylococci surviving in phagocytes. I know of only one paper dealing with a septic arthritis which did not respond to beta-lactam antibiotics and where staphylococci still were seen in aspirates of the arthritis, only intracellularly. Under those conditions, you could consider adding rifampicin because this has a better intracellular penetration.

Dr Kayser: Yes, I agree.

5. THE COLONIZATION RESISTANCE OF THE DIGESTIVE TRACT WITH SPECIAL EMPHASIS ON THE OROPHARYNX

D. VAN DER WAAIJ

In association with medical advances that have prolonged survival in both critically and chronically ill patients, including the introduction of a number of effective antimicrobial agents, a change in the pattern of many infections has occurred from acquisition outside of the hospital to acquisition within the hospital [1]. Bacterial infections of the respiratory tract have been reported to occur in 0.5 to 5.0% of hospitalized patients [2]. In a study of nosocomial infections in six community hospitals Eickhoff and associates [3] observed an adjusted rate of 3.5% for all types of nosocomial infections. Of these infections, 15.4% involved the respiratory tract and 53% of the pneumonias were associated with aerobic Gram-negative bacilli [3]. Bacteria may invade the alveolar level of the lung in sufficient numbers to produce infection by three routes: (1) haematogenously from a distant focus causing bacteremia such as Escherichia coli pneumonia during pyelonephritis [4]; (2) by suspension in inhaled gas which is a well recognized potential danger in the case of respiratory therapy, and (3) by aspiration from the pharynx, the most frequent route of lung infection. Indirect evidence supports this assumption [5, 6, 7], and the aspiration into the lung of radiopaque material instilled into the oropharynx of normal sleeping adults has been demonstrated [8]. Perhaps the most compelling evidence for the pharynx as a major source of infection of the bronchial tree is that from Johanson and coworkers, who studied the relationship of oropharynx colonization with Gram-negative bacilli to nosocomial pneumonia in patients admitted to a medical intensive care unit [9]. Ninety-five patients (45%) became colonized with Gram-negative bacteria. Nosocomial pneumonia developed in 26 patients, 22 of whom had previously been colonized with Gram-negatives. Thus nosocomial respiratory tract infections occurred in 23% of colonized patients but in only 3.3% of non-colonized patients.

EXPERIMENTAL RESEARCH ON MECHANISMS INVOLVED IN THE
CONTROL OF OROPHARYNX COLONIZATION

This brief survey of conditions leading to respiratory tract infections, showing
the importance of pharyngeal colonization, has made us decide to perform a
series of experiments on mice to study which factors control Gram-negative
pharyngeal colonization.

Oral contamination of groups of 20 mice with various Gram-negative bacilli
(E. coli, Klebs, pneumoniae, Ps. aeruginosa) in different doses per group were
performed. In each individual animal the presence of the contaminant in the
oropharyngeal and the faecal flora was studied quantitatively at 8-hourly intervals following contamination and after 24 hr every other day [10]. The results of
this study showed that contamination doses of 10^5 bacteria were rapidly cleared
from the digestive tract; the E. coli strain used persisted longest. The oropharyngeal cultures were negative for this contaminant in 18 of a group of 20 mice at 16
hr after contamination. However, it took several days before the faeces of the
majority of the mice in this group became negative. All animals were free at day
12 after contamination. When it was found that elimination of organisms from
the digestive tract depends on the number of orally administered cells, it was
decided to call this mechanism Colonization Resistance (CR). Higher contamination doses with 10^7 and 10^9 Gram-negative bacteria resulted in somewhat
longer lasting oropharyngeal colonization and more prolonged excretion in the
faeces of several weeks. As a result of incidental copophragy the evaluation of
oropharynx-clearance became unreliable after the second day following contamination so that no exact figures can be given of the high dose contamination
experiments. However, even contamination doses of 10^7 and 10^9 bacteria resulted in negative cultures at 8 and 16 hr after contamination in respectively 12 and 9
mice of the mice in each group of 20 animals. Following longer intervals of 24
and 48 hr copophragy may have caused 100% of the oropharyngeal cultures to
be positive again. From this observation one could tentatively conclude that in
mice persistent presence of Gram-negative bacteria in the oropharynx is secondary to faecal colonization.

IMMUNE SUPPRESSION AND ANTIBIOTIC TREATMENT

Treatment of mice with total body irradiation as well as with certain antibiotics
appeared to decrease the CR independently. In experiments of this kind, the CR
was found to result from cooperation between the host and the anaerobic
fraction of his anaerobic intestinal flora [10, 11].

Sublethal to lethal irradiation with 500–700 rads made the role of the host in
the CR mechanisms obvious. In these experiments the contribution of the gut

associated lymphoid tissue which excretes IgA was made likely as one of the important host factors [12]. However the contribution of mucosal damage by irradiation to prolonged pharyngeal colonization could not be excluded as it could not be independently studied. On the other hand, at the time after irradiation at which abnormal colonization was observed and at which IgA was not or minimally excreted (in the second and third weeks), no evidence of mucosal abnormality was seen by light microscopy in histological sections.

Antimicrobial treatment both topical in the gut aiming at gut sterilization [10] and systemic treatment [13] appeared to affect the CR negatively. Particularly during gut sterilization [10] but also during oral treatment with ampicillin [14], the CR decreased to an extremely low level. During such treatment oral contaminations, with doses as low as about 100 Gram-negative bacteria resistant to the antibiotic(s) used, 'take' and persist as long as antibiotic treatment is continued. If such antibiotic treatment is stopped in conventional environment allowing the animal to pick up normal flora constituents, or in case an aerobic intestinal microflora is implanted at the time the antibiotics are cleared from the intestines, the Gram-negative contaminant may be completely cleared from the digestive tract within two weeks 15].

On the basis of their influence on the CR, antimicrobial drugs have been

Table 1. Three classes of oral antimicrobial drugs regarding their effects on decrease of the colonization resistance (CR) in adult patients with normal intestinal absorption.

	Antimicrobial drugs	Normal daily oral dose in grams	Average CR-decreasing daily oral dose in grams	Estimated percentage absorption
Group I	Phenoxymethyl-pen	1.0–2.5	⩾ 0.6	70%
	Pheneticillin	1.0–3.0	⩾ 1.0	80%
	Cloxacillin	1.0–4.0	⩾ 1.0	80%
	Ampicillin	2.0–6.0	⩾ 1.0	80%
	Epicillin	2.0–6.0	⩾ 1.0	80%
	Tetracyclin	1.0–3.0	⩾ 1.0	80%
Group II	Amoxycillin	2.0–4.0	⩾ 2.3	90%
	Doxycyclin	0.1–0.2	⩾ 1.5	90%
Group III	Co-trimoxazole	2.0–4.0	⩾ 2.3	90%
	Cefradin	2.0–6.0	⩾ 9.0	90%
	Cefaclor	2.0–6.0	⩾ 9.0	90%
	Nalidixic acid	4.0–8.0	⩾40.0	90%
	Pivmecillinam	0.8–2.0	⩾ 9.0	90%

divided into three classes (Table 1): (1) antibiotics which decrease the CR at low doses, (2) antibiotics which decrease the CR only when given in high doses, and (3) antimicrobial drugs which do not affect the CR, even not when given at high doses [18]. If these latter drugs are given orally in sufficient doses they may more or less selectively eliminate Gram-negative bacilli from the entire digestive tract [16]. As mentioned above, disappearance of these bacteria from the oropharynx could be due secondarily to disappearance from the faeces. Treatment of patients with drugs which affect the CR even during low dose treatment, can have a dramatic effect on their oropharyngeal colonization pattern. In a pediatric ward with ten patients in which ampicillin was used for three weeks, 9 of 10 patients were pharyngeally colonized with often multi-resistant Gram-negative bacteria. At the start of treatment all had been negative.

In our irradiation experiments Strept. viridans, which according to some investigators [24] play a key-role in controlling Gram-negative colonization of the oropharynx, persisted in normal numbers at that site. In immunosuppressed individuals the contribution of these streptococci appears to be marginal or nonexistent.

EXPERIMENTAL CONSEQUENCES

The practical implications of the findings in animal experiments in which either antibiotics or immune suppression is achieved will be obvious. In this kind of experiment, in which both the contribution of the host and his microflora to the CR is affected, strict reversed isolation is indicated. To understand this one should recall that in such animals small dose contamination with a small number of (resistant) Gram-negative bacilli may very rapidly colonize many if not all individuals. This then may cause a lethal (mostly pulmonary) infection within 1 or 2 days after contamination.

CLINICAL ANALOGIES OF EXPERIMENTAL OBSERVATIONS

In man, as in small rodents, the oropharynx apparently does not provide a suitable environment for the growth of aerobic Gram-negative bacilli; only about 2–18% of normal persons harbor such organisms at any particular time [19, 20]. When multiple cultures are performed on healthy persons, the cumulative percentage of subjects with at least one positive culture increases, but previously positive persons are usually negative. This indicates that, as in mice, colonization is transient in normal persons. Furthermore, massive exposure of normal persons to these organisms does not result in colonization of the upper respiratory tract [21]. In order to define pharyngeal clearance mechanisms in healthy human subjects more fully, LaForce and coworkers undertook the following study which closely approximated our contamination experiments in mice. Normal volunteers gargled with suspensions of E. coli, K. pneumoniae and Proteus mirabilis, and the pharyngeal clearance of the microorganisms was subsequently followed [22]. With all three microorganisms studied, bacterial counts rapidly decreased. After three hours, less than 1% of the original inoculum was recovered. Extending the quantitative pharyngeal count curves of LaForce to more than 3 hr after contamination shows that very much as in mice, in man it also takes 8–16 hr before the contaminant is cleared totally. When radiolabelled E. coli was used in the challenge, the number of bacteria that were recovered decreased more rapidly than did the radio-active lable, suggesting that pharyngeal clearance mechanisms involve both physical clearance and bactericidal activity.

During the course of antibiotic treatment of pneumonia, aerobic and facultative Gram-negative rods may rapidly emerge as a significant part of the respiratory flora. In a study by Spencer and Philip [23], Escherichia coli, Klebsiella species or Pseudomonoas aeruginosa were found in the initial sputum cultures of 29% of patients hospitalized for pneumonia who had received antimicrobial agents before culture.

Although previous antimicrobial therapy can facilitate Gram-negative colonization in man, it may not be a prerequisite. Sprunt and coworkers [24] conclude from their observations in penicillin-treated patients that Gram-negative overgrowth is prevented in untreated individuals by the presence of Strept. viridans. They base their conclusion on the observation that overgrowth by E. coli was not seen in penicillin treated individuals who had resistant Strept. viridans in their oropharynx. However, they did not study the occurrence of other changes in the patients' microflora. It is conceivable, for example, that these patients had penicillin-resistant streptococci in addition to (other) resistant betalactamase producing bacteria in their intestines, and for that reason had an unchanged CR. Hofstra in our laboratory (unpublished data) has recently accumulated experimental evidence for this supposition. Furthermore, not only penicillins but also tetracyclins, certain cephalospirins and aminoglycosides, antibiotics which are often less active on Strept. viridans, are known to cause bacterial overgrowth in the pharynx in a number of cases. This makes the latter assumption of an effect on anaerobic bacteria in the intestines more plausible. The presence of resistant bacteria in the intestines that produce and liberate enzymes, such as betalactamases (penicillins and cephalosporins) and phospho- or acetyltransferases (aminoglycosides) which inactivate the antibiotic traces, that reach the gut through oral intake or by hepatic excretion, may better explain the conflicting literature concerning the occurrence of bacterial overgrowth during treatment with CR-decreasing antibiotics.

Almost completely in line with our experimental studies were observations recently described by LeFrock and coworkers [25]. They found in surgical patients who stayed for a minimum of three weeks in the hospital that in addition to an obvious effect of antibiotic treatment, a strong influence was noticeable in the duration of hospital confinement on the occurrence of Gram-negative colonization in the oropharynx. Furthermore, they confirmed the antibiotic-induced alteration of the normal Gram-negative oropharyngeal flora. This appeared, however, not to be a necessary precondition for the appearance of Gram-negative bacilli in this area. They related the appearance of Gram-negatives in the oropharynx with their presence in the faecal flora. In patients in whom Gram-negative bacilli appeared in the oropharynx, which were often of the same genera as the newly predominant faecal coliforms, a finding emerged which points to the latter as the origin of newly appearing oropharyngeal strains.

The conditions that promote oropharyngeal Gram-negative colonization as well as antimicrobial therapy are not known in detail. Stratford and co-workers noted however in 1968 [26] that the prevalence of colonization increased as the severity of illness increased. Similar results have been reported by Johanson and co-workers [27]. Indications of severity of illness associated with colonization were coma, hypotension, acidosis, azotemia, either marked leukocytosis of leukopenia and endotracheal intubation. Possibly the condition of the patient influences the condition of his oropharyngeal mucosa. 'Sick' mucosal cells may facilitate Gram-negative adherence and thereby pharyngeal colonization.

PREVENTION OF OROPHARYNGEAL GRAM-NEGATIVE COLONIZATION

Prevention of Gram-negative oropharyngeal colonization in patients who are likely to have 'sick' mucosal cells, i.e. patients with acute leukemia under chemotherapy, patients that were irradiated and patients with laryngectomy, has effectively been achieved by selective decontamination [28, 29]. Selective decontamination with antibiotics which do not decrease the CR could also be effective in other categories of patients who because of their condition are prone to get a broncho-pneumonia.

SUMMARY

Gram-negative lung infection appears often to be associated with and is perhaps secondarily to, Gram-negative pharynx colonization. Longlasting Gram-negative pharynx colonization is enhanced by treatment with antibiotics which decrease the colonization resistance of the digestive tract. This control mechanism of the normal colonization pattern appears to depend on cooperation between anaerobic intestinal bacteria and certain host factors. The latter may involve, among others, the gut associated lymphoid tissue and the condition of the mucosal cells. During hospitalization the chance for colonization of this area with Gram-negative micro-organisms is also enhanced in seriously ill patients without antimicrobial therapy.

REFERENCES

1. Rogers DE: The changing pattern of life threatening microbial disease. New Engl J Med 261:677, 1959.
2. Pierce AK, Sanford JP: Aearobic Gram-negative bacillary pneumonias. Amer Rev Resp Dis 110:647, 1974.
3. Eickhoff TC, Brachman PS, Bennett JV, Brown JF: Surveillance of nosocomial infections in community hospitals. I. Surveillance methods, effectiveness and initial results. J Inf Dis 120:305, 1969.
4. Tilloston JR, Lerner AM: Characteristics of pneumonias caused by Escherichia coli. New Engl J Med 277:115, 1967.
5. Lansing AM, Jamieson WG: Mechanisms of fever in pulmonary atelectasis. Arch Surg 87:168, 1963.
6. Kneeland Y, Price KM: Antibiotics and terminal pneumonia: A post-mortem microbiological study. Am J Med 29:967, 1960.
7. Mays BB, Thomas GD, Leonard JS, Southern PM, Pierce AK, Sanford JP: Gram-negative bacillary necrotizing pneumonia: A bacteriologic and histopathologic correlation. J Inf Dis 120:687, 1969.
8. Winfield JB, Sande MA, Gwaltney JM: Aspiration during sleep. JAMA 223:1288.
9. Johanson WG, Pierce AK, Thomas GD: Nosocomial respiratory infections with Gram-negative bacilli. Ann Intern Med 77:701.
10. Van der Waaij D, Berghuis-De Vries JM, Lekkerkerk-Van der Wees JEC: Colonization resistance of the digestive tract in conventional and antibiotic-treated mice. J. Hyg Camb 69: 405, 1971.
11. Van der Waaij D, Berghuis JM : Determination of the colonization resistance of the digestive tract in individual mice. J Hyg Camb 72:379, 1974.
12. Van der Waaij D, Heidt PJ: Intestinal bacterial ecology in relation to immunological factors and other defense mechanisms. In: Food and Immunology (Hambraeus L, Hanson LA, McFarlance H, eds.), Almqvist and Wiksell International Stockholm, 1977 p. 133.
13. Van der Waaij D, Berghuis JM, Lekkerkerk JEC: Colonization resistance of the degestive tract of mice during systemic antibiotic treatment. J. Hyg Camb 70:605, 1972.
14. Thijm HA, Van der Waaij D: The effect of three frequently applied antibiotics on the colonization resistance of the digestive tract of mice. J. Hyg Camb 82:397, 1979.
15. Van der Waaij D, Vossen JM, Korthals Altes C, Hartgrink C: Reconventionalization following antibiotic decontamination in man and animals. Amer J Clin Nurit 30:1887, 1977.
16. Van der Waaij D, Berghuis-De Vries JM: Selective elimination of Enterobacteriaceae species from the digestive tract in mice and monkeys. J Hyg Camb 72:205, 1974.
17. Boranić M, Van der Waaij D: The effect of the supply of oral antibiotic on the fecal flora and mortality of mouse radiation chimeras. J Inf Dis 122:83, 1970.
18. Van der Waaij D: Colonization resistance of the digestive tract as a major lead in the selection of antibiotics for therapy. In: New Criteria for antimicrobial therapy: mantenance of digestive tract CR. (Van der Waaij D, Verhoef J, Excerpta Medica, Amsterdam-Oxford, 1979, p. 271.

19. Johanson WG, Pierce AK, Sanford JP: Changing pharyngeal bacterial flora of hospitalized patients. Emergence of Gram-negative bacilli. New Engl J Med 281:1137, 1969.
20. Rosenthal S, Tager IB: Prevalence of Gram-negative rods in the normal pharyngeal flora. Ann Intern Med 83:355, 1975.
21. Meyers CE, James HA' Zippin C: The recovery of aerosolized bacteria from humans. I: Effects of varying exposure, sampling times, and subject variability. Arch Environ Health 2:384, 1961.
22. LaForce FM, Hopkins J, Trow R, Wang WLL: Human oral defense against Gram-negative rods. Am Rev Resp Dis 114:929, 1976.
23. Spencer RC, Philip JR: Effect of previous antimicrobial therapy on bacteriological findings in patients. with premary pneumonia. Lancet 2:349, 1973.
24. Sprunt K, Leidy GA, Redman W: Prevention of bacterial overgrowth. J Inf Dis, 123:1, 1971.
25. LeFrock JL, Ellis ChA, Weinstein L: The relation between aerobic fecal and oropharyngeal microflora in hospital patients. Amer J Med Sci 227:275, 1979.
26. Stratford B, Gallus AS, Matthiesson AM: Alteration of superficial bacterial flora in severely ill patients. Lancet 1:68, 1968.
27. Johanson WG, Pierce AK, Sanford JP: Changing pharyngeal bacterial flora of hospitalized patients: emergence of Gram-negative bacilli. New Engl J Med 281:1137, 1969.

DISCUSSION

Dr Kayser: I have a question relating to bacteriology. In examining the anaerobic flora of your animals, did you find a leading organism responsible for C.R.?

Dr Van der Waay: This is really the question. We have investigated this in co-operation with Nijmegen. Anaerobes isolated from humans were given stepwise to germ-free animals. It was observed that if about 50 different anaerobic species were given, the C.R. was back to normal. C.R. is apparently a cooperation among many different anaerobes. But if you ask which of all these species are more important than others, this is not yet known.

Dr Van der Meer: Up to now, we have been brought up with the division in broad-spectrum and narrow-spectrum antibiotics. You mention quite another division in those antibiotics namely those which affect colonization resistance and those which do not, and a group which is intermediate. If you look at your list, you see that a number of typical narrow-spectrum antibiotics like erythromycin belongs to Group I, the group that affects colonization resistance. Do you want to abolish the division broad-spectrum/narrow-spectrum? This is my first question. My second question is: should we avoid Group I antibiotics for patients in the hospital?

Dr Van der Waaij: The impact for hospital epidemiology is greater than one often thinks. For instance, in a pediatric ward where 10 patients were treated with ampicillin, we saw each week the gram-negative microflora expanding in these patients. These gram-negatives, often multi-resistant, showed up not only in the faecal flora, but unfortunately also in their oropharynx. Even though these children had a normal respiratory tract, it was regarded as a potential danger for contamination of other patients. This observation may explain why these days, we have in hospitals many more Enterobacter, Klebsiella and pseudomonas species, which are resistant to many antibiotics and are, so to speak, maintained in patients treated with Group I antibiotics. The patients colonized in this way do not necessarily suffer from this kind of treatment, but they still form a source of infection for the patient in the next bed. If the patient is in hospital for leukemia treatment, or tracheal intubation, or under other circumstances that decrease the condition of the mucosal lining in the oropharyngeal area, which in itself enhances gram-negative colonization, a 'source' in the next bed, may become a problem. So, for hospital epidemiological reasons, we prefer Group III and if necessary Group II antibiotics. If Group I antibiotics are used, the hospital epidemiologist should be informed if it is an I.T. ward, a hematological ward, or another ward where immune-compromised patients are present or to be expected.

Dr Van der Meer: Do you want to abandon the concept of broad and narrow spectrum?

Dr Van der Waaij: No because also among narrow-spectrum antibiotics some do affect the C.R. and others do not. For example, polymyxin-B and nalidixic acids are both examples of narrow-spectrum antimicrobial drugs. Although they are not at all absorbed, they will not affect the C.R. and therefore are not potentially dangerous.

Erythromycin on the other hand is C.R.-decreasing. This is presumably because clostridia species which play a role in the C.R. are suffering from erythromycin.

Dr Mattie: I think that before you really can state that the differences in effect on colonization resistance are really based on differences in absorption, this statement could be made more effective if a real correlation between the amount of antibiotic that is found in the faeces and the colonization resistance. For instance, one might guess that if those drugs of Group I are given parenterally, some of them will not be excreted in the gut, and therefore they should not influence colonization resistance. Some of them would be excreted by the liver in the gut and, in this indirect way, influence the colonization resistance. In this respect, I would expect, for instance, that bacampicillin would be just as bad as ampicillin.

The second point is that some of these drugs are conjugated by the liver into inactive forms. Nevertheless, those inactive forms reach the lower intestine hydrolyzed and the active drug again becomes free and could influence the colonization resistance. So, there are many hypotheses that can be tested to strengthen this concept.

A last question could be: are there any of these drugs that are excreted in the saliva, and do they–even if they have been administered parenterally–influence colonization resistance in the oropharyngeal region?

Dr Van der Waaij: I agree that we have certainly not investigated this in great detail for all antibiotics screened so far so that there remain many possible explanations for their effect on the C.R. like conjugation in the liver, as you said. However, co-trimoxazole has been investigated in more detail and several oral cephalosporins. Co-trimoxazole to begin with is certainly not completely absorbed and you may know that this is, nowadays, being used for selective decontamination–partial decontamination you call it. This means that you can give the drug in such a (high) dose that it will suppress the aerobic gram-negative flora to an undetectable concentration and yet not yet influence the C.R. If cotrimoxazole is given in higher doses than those given for the treatment of Pneumocystis carinii where doses as high as 9 and 12 tablets per day are being given, it does influence the C.R. During such high dose treatment certainly, there is a substantial effect on the C.R. Cefradin is in patients with liver cirrhosis not sufficiently absorbed after oral administration. Perhaps because of chronic enteritis, which these patients often have, they often show a very strong decrease of the C.R. during cefradin administration. This is also the case in patients with blind loops, which for this reason have reduced intestinal absorption. In patients with no known abnormalities in their intestinal absorption we have not found measurable cefradin concentrations in the faecal material. This does not mean that it is not here. It is known of cephalosporins that they are inactivated by fecal material. That means that the complexes (compounds) that are in the colon are apparently not biologically active and therefore may explain the absence of an effect on the CR and the absence of measurable concentrations in the fecal material.

On the other hand carbenicillin as a systemic drug is amazingly unfriendly to the flora. If it is given in a dose of 30 grams i.v. a day, you may have a patient with virtually bacteria-free faeces. Just make a gram smear of the faeces of such a patient, and you will see that it is often sterile.

Dr Mouton: I would like to ask Professor Van der Waaij what his opinion is with regard to the implications for the choice in therapy. If he accepts this trinity of groups of antibiotics, then to me it does not imply that one has to choose from the group that gives the least resistance decrease. I would think there would be other important parameters in the choise of an antibiotic. For instance, if we look at the cephalosporins, where you find that cefradin is in Group III–the other cephalosporins are not mentioned here, but I expect you will find them in category I or II–this does not imply that one has to choose cefradin because it may be much less active. If you are treating patients for an infection, then I think the drug which does the best job should be chosen.

I would like to have your opinion especially with regard to mesicillin and cefradin, both of which have disadvantages in this respect.

Dr Van der Waaij: I certainly agree that criterium number one for the selection of an antibiotic for treatment of an infection is the sensitivity of the bacterium that causes the infection to the drug. That is without question.

The point is however, that if in a particular case you have to use antibiotics of Group I, then you should realize what you are doing. If you are treating a patient in a psychiatric ward, or a dermatological ward, I would not hesitate to use it because, epidemiologically, it will not be of any potential harm to any of the other patients. If it does any harm, in terms of overgrowth by resistant bacteria or yeasts it may only be of potential danger to the patient him/herself.

However, as I mentioned, in an environment where patients with decreased resistance to infection or who have an altered mucosal sensitivity like in elderly patients, or who have a viral infection of the upper respiratory tract, to enhance colonization by gram-negative, including haemophilus, if those kinds of patients are around I would think twice. If application of Group III antibiotics is not possible, one should take precautions. One could take isolation precautions or apply selective decontamination in addition to such therapy, i.e. steps required to limit the spread of infection.

Dr Mouton: You just mentioned sensitivity. That is, of course, important, but more important is experience in treatment of patients in which you find that one drug is better than the other. Even if you know that the colonization resistance may be adversely affected when you use another drug from Group III, for instance, then I would choose the one which does the job in preference to one which does it less so. I am somewhat afraid that this is going to lead us on a line without regard to the effectiveness of the antibiotics themselves.

Dr Van der Waaij: I repeat, I completely agree. If you have to treat a patient with a severe infection, then nobody would argue that you should use the most potent and effective cephalosporin. Obviously, you must use the best drug as soon as possible.

However, in these patients, we can often prevent infection. Take burn cases. If you know beforehand that a patient is going to be extremely susceptible to infection, why not make use of this information and eliminate the potential pathogens before infection takes place. This is what we do with leukemic patients, in liver-transplant and kidney-transplant patients, in burns, and that we may use in bone-marrow transplant patients in the future.

Dr Mouton: This is quite another area, when you talk about prevention of colonization and infection in patients with impaired host resistance. I was just talking about the patient in the hospital with an infection. Should one take into account your data or not?

Dr Van der Waaij: If you can, I would. We do. If you cannot, then you obviously have to make use of another antibiotic. C.R.-saving is not a religion, but a practical approach in antibiotic therapy. The point is know what you are doing to the flora of the patient which is not harmful.

Dr Davies: I was very interested to hear what you have to say. Have you got really hard evidence that, if you follow this policy of yours, you actually are cutting down the cross-infection that you talk about. It sounds very nice in theory, but is it really happening in practise? That is the first question.

The second question is: clearly, in those three groups, we have in chronic bronchitis for example 3 drugs which are very widely used–erythromycin in Group I, amoxycillin in Group II and co-trimoxazole in Group III. Apart from the reasons you have given us for not using antibiotics in Group III, can you actually–outside of the intensive-care unit–say that they have other effects in the ordinary chronic bronchitis, for example, in terms of gastrointestinal upset, untoward effects? Do they really upset the gut? That would be another reason why we should choose the better absorbed antibiotics.

Dr. Van der Waaij: In an IT-ward, we have often seen in the past that, if one patient came in with multi-resistant Enterobacter or Klebsiella, it tended to spread quite rapidly among the other patients in the same ward. This was seldomly correlated with whether or not those patients were treated with antibiotics. However, these patients had all undergone surgery of long duration; i.e. they had been under anesthesia for quite a long time; furthermore they were sometimes unconscious. Anyway, they all had underlying conditions which are known, from the literature, to enhance the colonization of the oropharyngeal area. So, if you have patients who have an altered 'mucosal susceptibility' to gram-negative colonization, in a ward like an IT I would avoid introducing or facilitating the presence of multiresistant bacteria among patients in that ward. A patient who comes in with multi-resistant Enterobacter species or Klebsiella species and who is treated with gentamicin and carbenicillin for example because these are the only two drugs to which the strain in question is sensitive, then one may maintain in the intestinal tract of those patients this particular bacterium. At this site it may become resistant and cause overgrowth in the alimentary tract. From there it may spread via the nurses' hands–although it should not–to other patients. We feel that one should take into consideration how heavily a patient is or may become colonized by gram-negatives. The chance for overgrowth is certainly enhanced by the use of Group I antibiotics. Once you know it, put the patient in a separate room and take precautions, like additional oral treatment with polymyxin B or other drugs for selective decontamination. This will prevent development of resistance and overgrowth, so that the patient in question will not be a potential danger.

Your second question was?

Dr Davies: Some people in England say that perhaps the reason that amoxycillin in some people's hands seems to produce few gastrointestinal upsets, compared to ampicillin, might be because it is better absorbed. Can you see differences in untoward effects, such as gastrointestinal disturbances, between the drugs in the three groups, based on your theoretical reasons?

Dr Van der Waaij: A decrease of the C.R. is more often seen during treatment with ampicillin than during amoxycillin therapy which could be due to better absorption of the latter drug.

Dr Davies: The only reason I am going about this is because it seems to us, as clinicians, that there is a very wide choice of antibiotics to use; so, particularly for chronic bronchitis. Clearly one is looking for drugs that cost less, but also for drugs which will give less untoward effects, as becoming a real, important issue to us–or should be.

Dr Kunst: I would like to make two comments. I think when you are talking about treatment of a moderately or severely ill patient for infection and when absorption of the antibiotic plays a role–especially in that kind of patient–I think that absorption is something you cannot predict. It might be that the absorption, with an impaired gastrointestinal function in the ill patient, is rather bad. So, there would not be a reason to take one of the absorbable Group III drugs for treatment in that respect as well.

The second comment I would like to make: I am interested in antibiotic treatment in burns. Theoretically, although prophylaxis in burns is a rather difficult matter, one prefers, when giving prophylaxis, to choose Group III antibiotics. I had bad experiences with several infected patients from Italy, where they give routinely cephalosporins in all their burn patients. I think the problem of multi-resistant strains (prophylaxis may not be the main factor in the development of multi-resistant strains), especially in that country where cephalosporin prophylaxis is routinely given, is a very difficult one. So, in practise, the prophylaxis of the absorbable Group III drugs does not work as well as you might expect from you colony-resistance problem.

Dr Kerrebijn: I have two questions. How long does it take for the colonization resistance to return to its normal level after stopping the antibiotic treatment? The second question is: can you speculate a little about the impact of your theory on the treatment of chronic pseudomonas infections in the lungs, like in cystic fibrosis for instance, with carbenicillin, gentamicin or with other antibiotics.

Dr Van der Waaij: Louria has indicated that it takes about 14 days for the colonization pattern of the throat to return to normal. In our hands, it takes usually a little shorter. It is more in the order of one week than two weeks, before it returns back to normal.

In answer to your question regarding the pseudomonas infections in Groningen and as far as I know in Utrecht at the moment such patients are treated orally with polymicine-B, which is not absorbed, thereby eliminating the pseudomonas from the oropharynx and intestines of those patients. By preventing continuous reinfection of the lungs a better outcome of the treatment of pseudomonas infections may be achieved. Also, the remissions that are obtained in those patients are much longer. When they are maintained on oral polyxin it may take even months before they have a relapse with pseudomonas.

Here again is an indication that if one keeps the oropharyngeal area of these patients 'clean'–and this is often secondary to the faecal colonization pattern, in one way or another–the results are much better when therapy without selective decontamination treatment is applied.

Dr Kayser: I just want to make a comment with regard to your third question, Dr Davies. Would not one infection of the gastrointestinal tract be connected with the use of some of the Group I drugs, i.e., the antibiotic-induced enterocolitis? pseudomembranous entero-colitis?

Dr Van der Waaij: You asked me whether pseudomembranous colitis occurs with Group III antibiotics? It has been described for cefradin but as far as I know not with co-trimoxazole.

Dr Mattie: I want to come back to the point Dr Davies raised about intestinal discomfort. Decrease of colonization resistance depends on poor absorbability. On the other hand, very probably, intestinal discomfort depends on poor absorbability of this kind of drug. Does that not imply that intestinal discomfort is caused by changes in bacterial flora? You seem to imply that with your question.

Dr Davies: Can I just comment? I was really asking for information. I was just suggesting it. I was really interested to see whether our speaker had got any suggestions. Whether he had noticed changes, or had anybody looked at these sorts of changes and related them to the side effects you were mentioning?

Dr Van der Waaij: The group in Kentucky has investigated that in an animal model and it appears that the anaerobes do break down the mucus that is excreted in the colon. The intestinal mucus is apparently hypertonic, and in that way keeps water in the colon unabsorbed. This was their explanation for the fact that these animals also have diarrhea during treatment with antibiotics of the kind which do inhibit the breakdown of the mucus in the colon. But, obviously, this is only part of the picture.

Dr Van Boven: I should like to ask two questions. In the light of the short prophylactic use of antibiotics–the 24-hour duration of antibiotics in surgery–how fast can the colonization resistance be reduced by antibiotic use? The second question: I can see the relevance of this classification of antibiotics for use in hospitals, but what is the relevance for general practise?

Dr Van der Waaij: At the moment, I do not see any relevance for the general practitioner. Here, you are dealing with one patient in a relatively healthy environment. This morning, we already heard that colonization with resistant bacteria in the outpatient is certainly less than it is in the hospital patient.

With regard to your first question concerning the one-day treatment, even Louria in his most dramatic paper, which I quoted this morning, found an interval of only two days. In our experience, it is usually somewhat longer and Klastersky published two years ago, that bacterial overgrowth was noted when patients were treated with broad-spectrum antibiotics for longer than two weeks. This means that the effect on the C.R. very much depends on the dose given and the route of administration. If given orally, it will have a much more rapid effect than following intravenous administration and in a moderate dose.

PHARMACOKINETICS

6. GENERAL REVIEW ON PHARMACOKINETICS OF ANTIMICROBIAL DRUGS IN RELATION TO RESPIRATORY INFECTIONS

H. MATTIE

THE TWO-COMPARTMENT PHARMACOKINETIC MODEL

Even those who are only slightly familiar with pharmacokinetic concepts will know that the two-compartment open model is often used in explaining the course of plasma concentrations of a drug [1]. What is often forgotten is that this model is primarily not a biological model but a mathematical one, although the latter is compatible with some biological facts. This has led to the assumption that a biological model is the basis for the mathematical model.

The biological model implies that the drug, once it enters the system, is instantaneously and homogeneously distributed over a certain part of the body, from which it is distributed further to a second part of the body. The drug moves freely between these so-called compartments, but is removed from the body by the first, or central, compartment only. The rates of distribution and elimination are all concentration dependent. For elimination it is compatible with the biological concept of clearance, for distribution it seems to be compatible with the concept of passive diffusion.

These concepts lead to a set of mathematical equations, regarding the rate of change of concentrations in the first volume of distribution (dC_1/dt) and in the second volume of distribution (dC_2/dt):

$$dC_1/dt = -k_{12}C_1 - k_eC_1 + k_{21}C_2 \tag{1}$$

$$dC_2/dt = k_{12}C_1 - k_{21}C_2 \tag{2}$$

leading to

$$C_1 = Ae^{-\alpha t} + Be^{-\beta t} \tag{3}$$

in which A, B, α and β are shorthand for rather complicated equations made up from k_{12}, k_{21}, k_e, administered dose (D) and volumes of distribution (V_1 and V_2).

In many instances experimental data are easy to fit, graphically or by computer, to this so called bi-exponential curve. From the thus derived values for A, B, α and β, and the dose, the values for k_{12}, k_{21}, k_e, V_1 and V_2 can of course be calculated. But this in no way *proves* that our assumptions regarding the biological model are true.

Even so, by giving names to parameters like k and V, namely *distribution constants, elimination constants* and *volumes of distribution* of *central* and *peripheral compartment* and using those over and over again, many authors seem to believe that pharmacokinetic models regarding a drug are a biological reality, instead of realizing that the drug concentration in the plasma only behaves *as if* the body was made up in the way described above. It should be emphasized, however, that the good fit of experimental data to equation (3) only applies for plasma concentrations, where the plasma sample is considered a sample from the first or 'central' compartment. This is already a doubtful assumption, because many drugs are protein bound, the calculated value for V_1 is more often greater than the plasma volume and the protein content of the extravascular fluid is lower than that in plasma. From a biological point of view it must therefore be clear that the 2-compartment open model leads not to truly quantative biological parameters, but only to useful descriptions of the course of plasma concentrations.

TISSUE PHARMACOKINETICS

The considerations given above have to be kept in mind when samples are being taken from others parts of the body than the plasma. The tissues sampled in this direct manner, have nothing to do with the second volume of distribution, often erroneously called 'tissue compartment'.

To describe drug movements between the blood and a particular tissue or body fluid other mathematical models should be developed. Only if the tissue compartment under investigation is identical to the second volume of distribution do equations (1) and (2) apply. This, however, is seldom the case. In recent literature on tissue concentrations of antibiotics the complexity of the problem is often overlooked leading to disagreement or apparent contradiction. One of the first questions to be raised is that of the definition of tissue concentration. Very often tissue concentration is defined as the drug content of a certain volume of homogenized tissue. We should realize that most tissues are histologically not homogeneous at all. At least we should consider them as consisting of a cellular and an extracellular component. Of course, it depends on the kind of drug, but for many antibiotics the intracellular concentration is not relevant in relation to the site of infection. Moreover, just as the protein bound part of drug in plasma is understood not to contribute to the effective concentration, neither does the amount of drug that is bound to tissue components, e.g. bone, contribute to its efficacy. What matters is the freely available concentration at the site of infection.

A different approach has been to collect true extracellular fluid and determine its antibiotic concentrations. Some data are available on fluid expressed from tissue under high pressure [2], and on lymph from peripheral lymph ducts [3, 4], but much recent data on tissue concentrations is derived from extracellular fluid collected in reservoirs of some volume [5]. What is overlooked in many publications is that the content of such a reservoir is not at all comparable to any extracellular fluid from a pharmacokinetic viewpoint, even if the chemical composition is often similar [6]. It should be evident that a foreign compound like a drug enters into such a reservoir by simple diffusion, and accordingly should obey the laws of diffusion, or Fick's law [7, 8]. There will also be a concentration gradient from the periphery of the reservoir to its centre, the peripheral concentration being not much less than that outside the reservoir, while it will take some time before any measurable concentration will exist at all at the centre [7].

With this in mind and looking at the available data, one often arrives at different conclusions than the authors themselves. Data on homogenized tissues prove that equilibration between plasma and tissue is often a rapid process, contrary to what authors working with tissue reservoirs conclude. On the other hand, some data provided by those authors permit the calculation of real diffusion constants, according to Fick's law, showing that this theory holds, and that diffusion in itself is not always a slow process [7].

PHARMACOKINETIC CONSIDERATIONS REGARDING THE RESPIRATORY SYSTEM

Turning now to antibiotic concentrations at the site of infections of the respiratory system, it is clear that this system is pharmacokinetically not homogeneous. Sinusal mucosa and tonsils are typical tissues that will be reached with more or less delay from the blood stream over a diffusion boundary. After intravenous injection of the drug, the non-protein bound part will diffuse rapidly into the tissue, according to the concentration gradient. At some time point the declining (free) plasma concentrations will equal rising (free) tissue-water concentrations. After that moment the concentration gradient reverses and the drug diffuses back to the blood stream. Therefore the tissue to plasma concentration ratio, which is often used to express the diffusability of the drug, is not a good parameter. First of all it changes in time after administration, and further it depends on the kinetics of the drug in plasma: the same diffusability leads to more favourable tissue to plasma ratios the lesser the plasma clearance is, approaching a ratio of 1, if there should be no plasma clearance at all.

It goes without saying that this ratio refers to protein free drug in plasma water and extracellular tissue water, respectively. Antibiotics, like most β-lactam antibiotics and the aminoglycosides that virtually do not enter cells, will always show ratios less than 1 if whole tissue content is determined.

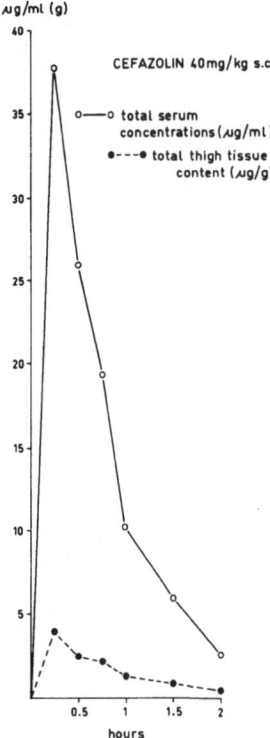

Fig. 1. Concentration time curve of cefazolin in mice. (Each point represents the mean value of two mice).

If it is possible to follow plasma concentrations and tissue contents in time, then the rate of diffusion may be determined. An example is shown in Fig. 1 regarding plasma concentrations and muscle tissue contents of cefazolin in mice [9]. Probably a good approximation of the relation between both curves is

$$C_T = k_{PT} \times AUC_P - k_{TP} \times AUC_T$$

in which C_T is tissue content, k_{PT} and k_{TP} the rate of diffusion from plasma to tissue and vice versa, and AUC_P and AUC_T the areas under the curves of plasma concentrations and tissue contents respectively. Of course, for free plasma concentrations and free extracellular tissue concentrations k_{PT} is equal to k_{TP}.

If one follows the whole concentration course until C_T becomes 0 it is possible to calculate the ratio k_{PT}/k_{TP}. The value thus found, 0.22, corresponds well with the extracellular volume. If one corrects for this (Fig. 2) it turns out that diffusion is indeed very rapid.

To interpret concentrations in bronchial secretions properly is far more difficult. First of all it should be kept in mind that, for example, after a bolus

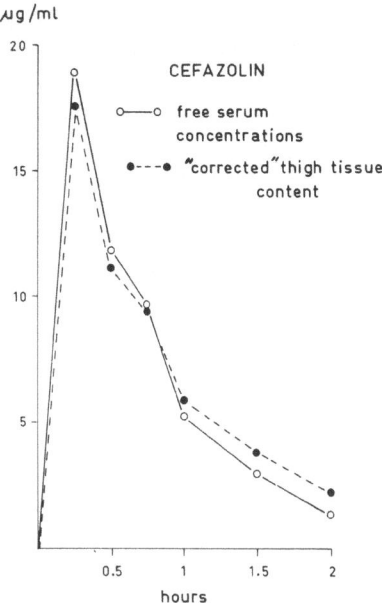

μg/ml

Fig. 2. Concentration-time curves of cefazolin in mice. The values of Fig. 1 are corrected for protein binding in plasma and for calculated interstitial concentrations in muscle, respectively.

injection the total amount of drug first passes via the circulation through the lungs at very high concentrations. It is conceivable that a greater or lesser part is absorbed by the lung tissue from where it is available for diffusion into the bronchial mucosa [10]. In the literature on this subject this possibility is very often overlooked. It might explain the rapid initial rise in bronchial mucus concentrations that is sometimes seen. The further passage from the circulation to the bronchial mucosa will probably not differ very much from that described above for other tissues. But thereafter it diffuses into the bronchial secretions, probably following Fick's law of diffusion into a large space, where a concentration gradient exists within this space.

The data already mentioned on concentrations in tissue reservoirs should probably be interpreted in this way. The data published by Barza et al. [11] regarding diffusion into fibrin clots do indeed fit this conception [7].

In the literature on concentrations in bronchial secretions it is not always clear what is really sampled. If it is sputum, does the produced amount contain all the drug that has passed into the bronchial secretion during the stated time interval? In this case it would be better to calculate the total amount of drug instead of the concentration, for the same reason that urinary concentrations do not give any information on renal clearance. If on the other hand small samples are taken by fibrescope, then the situation may be more like sampling from a tissue reservoir.

An additional problem is that bronchial secretions are constantly produced. For drugs that do not enter into the mucus-producing cells this implies that the drug in the bronchial secretion is constantly diluted. This might explain why, for example, aminoglycoside concentrations in bronchial secretions already decline when they are still much lower than plasma concentration [12]. By diffusion alone this could not be explained and active removal of the drug looks very improbable. If dilution is indeed the explanation, then the concentration ratio mucus to plasma is determined by the relation between rate of diffusion and rate of bronchial secretion.

EFFECTIVE CONCENTRATIONS

Meanwhile, the reason that so many authors who collect data on concentrations of antibiotic in bronchial secretions, sinusal mucosa etc., do not bother about the pharmacokinetic model, seems to be that their main interest is in the end result, namely the concentrations themselves. Those concentrations are then directly compared with the minimal inhibitory concentration (MIC) or bactericidal concentration (MBC) in vitro to establish their efficacy, or to compare the efficacy of different dosage regimens. This practice is, however, not warranted by any experimental proof, nor by data on the efficacy of different kinds of antibiotics in bronchial secretions. On the other hand it is well known that the efficacy in vitro greatly depends on the medium. Therefore the MIC in vitro should not be applied to concentrations in vivo.

Last but not least, the technique of measurement of concentrations should be mentioned. Just because the antibacterial activity of antibiotics often greatly depends on the medium, the utmost care should be taken that in this respect samples and standards should be similar. It is therefore amazing that some authors use saline for standards, without mentioning whether this is suitable or not, and others use serum, again without much justification. Without this provision the obtained results deserve to be regarded with a critical eye.

CONCLUSION

The existing pharmacokinetic models are often not applicable to determine the degree of diffusability into tissues or larger volumes of body fluids. With some precautions, and the proper interpretation of data in the light of biological reality, it should be possible to provide more insight in the penetratory properties of antibiotics. Measuring concentrations only for their absolute value does not provide much information on the efficacy of antibacterial drugs.

REFERENCES

1. Curry SH: Drug disposition and pharmacokinetics with a consideration of pharmacological and clinical relationships. Blackwell Scientific Publications, Oxford, London, Edinburgh, Melbourne 1974.

2. Mascherpa P: La méthode de l'expression fractionnée des organes et ses applications en chimie biologique et en pharmacodynamic. Bull Sté Chim Biol 33:1282, 1951.

3. Verweij WF, Williams Jr, Hr: Relationship between the concentrations of various penicillins in plasma and peripheral lymph. In: Antimicr Agents and Chemother 1962: Proc 2nd Interscience Conf (Hobby GL, ed.), Amer. Soc. Microbiol., Bethesda, 1963, p. 476

4. Acred P, Brown DM, Clark BF, Mizen L: The distribution of antibacterial agents between plasma and lymph in the dog. Brit J Pharmacol 39:439, 1970.

5. Chisholm GD: The tissue cage model in the distribution of antibacterial agents. Scand J Inf Dis, Suppl 14:118, 1978.

6. Eichenberg HU: What is interstitial fluid? Scand J Inf Dis, Suppl 14:166, 1978.

7. Mattie H, Kunst MW: The in vivo significance of tissue concentrations. Infection, Suppl 2 (4):164, 1976.

8. Bergan T: Kinetics of tissue penetrations. Scand J Inf Dis, Suppl 14:36, 1978

9. Kunst MW, Mattie H: Cefazolin and cephradin: relationship between serum concentrations and tissue contents in mice. Infection 6:166, 1978.

10. Hori R, Yoshida H, Okumura K: Tissue distribution and metabolism of drugs. IV: Accumulation and penetration of some antibiotics in rat lungs. Chem Pharm Bull 27 (6):1321, 1979.

11. Barza M, Weinstein L: Penetration of antibiotics into fibrin loci in vivo. Comparison of penetration of ampicillin into fibrin clots, abscesses, and 'interstitial fluid'. J Inf Dis, 129:59, 1974.

12. Pennington J: Kinetics of penetration and clearance of antibiotics in respiratory secretions. In: Immunologic and Infectious Reactions in the Lung. Kirkpatrick ChH, Reynolds HY (eds.), 1976, p. 355.

DISCUSSION

Dr Michel: This may be a question of semantics, but have you strong feelings about a two-compartment model? Recently, in the literature concerning aminoglycosides, there was an attempt to explain that for this case we should preferably use a three-compartment model.

Dr Mattie: Well, actually the body is not so constituted that it only consists of two compartments; there are several compartments. From a mathematical point of view, to describe the plasma concentration curves, a two-compartment model often suffices. Sometimes, it does not and you should apply a three-compartment model. But again, these compartments often have nothing to do with the actual compartment from which samples are taken. The compartment that you are actually sampling is one of the many real compartments, with its own diffusion characteristics, while diffusion coefficients calculated from plasma concentrations are just overall parameters. For instance, the diffusion from the plasma to the middle ear is not reflected at all in the plasma concentrations, of course, because the compartment is much too small.

Dr Michel: But the enormous binding of aminoglycosides in the kidney does show up in a careful estimation of the plasma. For years and years, we did not realize this, because we did not take the trouble to look long enough at the plasma.

Dr Mattie: That is a particular case in which active uptake is engaged and the ordinary pharmacokinetic models do not apply at all.

Dr Michel: Indeed, it is a particular case for a large group of antibiotics.

Chairman: I think we will come back to this later on, in a more general discussion.

7. PENETRATION OF MACROLIDES INTO THE RESPIRATORY TRACT

F. FRASCHINI, M. FALCHI and V. COPPONI

Almost all antibiotics are known to appear in saliva and bronchial secretions, at least in trace amounts. However, in these secretions certain antibiotics reach concentrations equivalent to or even higher than the M.I.C. for numerous bacterial strains or genera that are responsible for the most common respiratory disorders. Such pharmacokinetic data are therefore of great interest for the choice of the most suitable antibiotic therapy for a bacterial infection of the respiratory system. The exact mechanism underlying penetration of antibiotics into the lung, of their secretion into saliva, or at least all of the factors involved, are still unknown [1]. Antibiotics reach the bronchopulmonary system and saliva in two completely different ways. Salivary secretion is related to pharmacokinetic processes and to the degree of tropism for tissues, because the salivary glands do not offer any appreciable barrier to antibiotic penetration. On the contrary, the concentration of antibiotics in the respiratory system seems to be completely independent of serum levels, and in healthy subjects bronchopulmonary concentrations are usually much lower (30–40 times) than those in the serum. These pharmacokinetic data led to the conclusion that there must be a blood-lung, or rather a blood-bronchi, barrier that may be even less permeable than the better-known blood-brain barrier [2]. However, certain antibiotics reach surprisingly high (macrolides, lincomycin, rifampin, chloramphenicol) or at least therapeutic (ampicillin, cefalosporins) levels in the bronchi. This phenomenon has led some authors to suggest the existence of a mechanism of active transport able to concentrate some antibiotics selectively in bronchial mucus [3,4].

Generally, there is a correlation between the molecular weight and the penetration of an antibiotic into the bronchopulmonary system, high molecular weight corresponding to good penetration; for instance, cloxacillin reaches levels twice as high as those of ampicillin and the concentrations of tetracyclins, lincomycin, fusidic acid, novobiocin, rifampin, and erythromycin are even higher. Chloramphenicol and tiamphenicol are exceptions, because their small molecular size enables them to diffuse rapidly everywhere [1, 4, 5]. A similar correlation between molecular weight and penetration has been reported for the biliary secretion of antibiotics. Diffusion into bronchi is influenced by other factors as well, such as solubility in lipids and binding to serum proteins. These observations lead to the conclusion that there is hardly ever any correlation between the serum level of antibiotic and its concentration in the respiratory system. For this reason the most reliable parameter for the evaluation of the efficacy and clinical outcome of a given form of treatment for respiratory infections is exact determination of the tropism of the antibiotic in question for the bronchopulmonary system.

Since antibiotic concentrations in the saliva and the bronchopulmonary system generally show correlation, it should be possible to extrapolate data on tropism for the respiratory system simply by examining salivary levels. However, this relationship is only sufficiently linear for a few antibiotics and may be applied only for macrolides, which according to some authors show a certain parallelism between the serum and salivary levels (erythromycin) [6]. However, other studies have confirmed the finding of excellent salivary concentrations reached by macrolides (erythromycin and spiramycin) [7], but have also shown that at a low dosage (single doses of 500 mg per os) penetration into the saliva is rather erratic, being completely absent in 35% of the (healthy) subjects examined; moreover, similar findings have been made with ampicillin. Therefore, salivary tropism of antibiotics is a decidedly subjective phenomenon, and high serum levels do not necessarily correspond with high salivary levels or vice versa. Furthermore, in the case of ampicillin elevated salivary concentrations do not always accompany to high concentrations in the sputum, and an increase of the dosage does not always result in a linear rise of the salivary levels. In fact, the dose must be at least quadrupled to obtain a significant increase in the salivary levels [8].

Therefore, such levels alone do not offer a reliable parameter for the determination of the tropism of an antibiotic for the respiratory system. The tropism of an antibiotic for the salivary glands is a very favourable characteristic in cases of dental infections or infections of the upper respiratory tract; in fact, macrolides are excreted into saliva at concentrations lying in the range of serum levels, and are therefore considered antibiotics of first choice (also because of their antimicrobial spectrum) in dentistry and otorhinolaryngology.

The pharmacokinetic data that permit accurate evaluation of the tropism of chemotherapeutic agents for the lower respiratory tract concern their concentration in bronchial mucus or in sputum. It is of great interest that the kinetics shown by antibiotics in the respiratory system are similar to those related to the passage of the blood-brain barrier. The penetration of chemotherapeutic agents into the central nervous system is generally proportional to the inflammatory state of the meninges. Similarly, the concentration of an antibiotic in bronchial fluid reaches levels proportional to the purulence of such secretion. The relationship between the concentration of an antibiotic and the degree of purulence is not, however, strictly linear, because at low doses the dependence of concentration on pus is hardly ever significant. Only at high doses is the increase of the antibiotic concentration in bronchial secretion proportional to the quantity of pus and to the degree of inflammation in the respiratory system [8].

On the whole, and taking the variability of antibiotic kinetics in the broncho-pulmonary system into account, the finding of therapeutic levels (approaching or superior to M.I.C. values) in the sputum may be considered a reliable index for evaluation of the ultimate clinical efficacy of an antibiotic used to treat bacterial infections of the respiratory system. However, when antibiotic levels in the sputum are lower than the M.I.C., this does not mean that therapy will certainly be ineffective, because the clinical outcome has been favourable in many cases despite low sputum levels. This phenomenon may be ascribed to the fact that an antibiotic presenting low concentrations in bronchial secretions as a whole, actually reaches much higher concentrations in the individual bronchioli, and these concentrations are more than sufficient to inhibit the infecting micro-organism. Moreover, penetration of antibiotics into the respiratory system differs between bronchioli and is strictly proportional to the degree of inflammation in the individual bronchiole [1].

In sum, when an antibiotic shows positive tropism for the broncho-pulmonary system it tends to accumulate, because of the decrease of the blood-bronchi barrier by the local inflammation, at the sites damaged the most by the bacterial infection, and thus reaches levels in the bronchial fluid that are much higher than the average concentrations found in the sputum. As already mentioned, penetration into the bronchopulmonary system differs considerably from one antibiotic to another. In this respect, macrolides show almost ideal pharmacokinetics [9]. The antimicrobial spectrum, absorption, distribution in the organism, and excretion are very similar for all macrolides and do not differ substantially from those of the leader of this group of antibiotics, erythromycin, which, owing to its very favourable characteristics, is considered the ideal chemotherapeutic agent for the treatment of the most common respiratory infections. The antibacterial spectrum of erythromycin includes the micro-organisms most frequently responsible for the majority of bronchopulmonary infectious disorders, since this drug is active against Gram-positive bacteria and Haemophilus influenzae, which have been shown statistically to be the most frequent causative agents in respiratory pathology [10, 11]. From the clinical point of view erythromycin is one of the most reliable antibiotics, and the results reported in the course of several decades have been constantly satisfactory. The pharmacokinetic data reported for erythromycin in the respiratory system are very favourable [2, 6, 12]. Furthermore, in the pulmonary parenchyme this antibiotic reaches higher concentrations than those found in the serum, which constitutes proof of a particular tropism for the respiratory system.

In respect to this point, good distribution of erythromycin was observed as early as 1961 [13] in a certain number of cases of ischaemic lungs, where measurable levels were present in both proximal and distal parts of the organ even two days after antibiotic therapy had been instituted. In a recent study [14], subjects who had undergone lung removal, necessitated by neoplasia, showed excellent pulmonary levels of erythromycin, the concentrations being slightly below 5 μg/ml. All of these patients had received the antibiotic orally during the two days prior to the operation (1.5 g per day), plus a final single dose of 500 mg per os given $3\frac{1}{2}$ hr before surgery. The levels of erythromycin in the bronchial secretions of the excised lung were equivalent to 3 μg/ml.

These findings indicate, or rather confirm, that erythromycin shows particular kinetics in the respiratory system, i.e., it accumulates selectively in pulmonary tissue and then passes into bronchial secretions, not only still in an active form but also at levels that closely approach maximum serum concentrations [14]. These high bronchial concentrations permit erythromycin, which is bacteriostatic at minimum inhibitory concentrations, to become absolutely bactericidal, as shown by a study in which the bacteria were collected by means of bronchial endoscopy and examined electronmicroscopically [15]. Erythromycin was also found to reach efficacious therapeutic levels in the sputum of a series of patients suffering from chronic respiratory infections who were being treated with 1.25 to 2.0 g of the antibiotic daily per os. Starting in the 7th hr after administration, the erythromycin level remained between 1 and 1.75 μg/ml sputum during the next 24 hr [2].

No studies on other antibiotics belonging to the macrolide group have been performed to determine the concentrations obtainable in pulmonary tissue and bronchial secretions.

Erythromycin diffuses rapidly into the saliva, but levels fall at the same rate as those in the serum and on the whole prove to be lower than the latter. In the period between 0.5 and 6 hr after a single oral dose of 500 mg, average salivary levels remain at about 10% of the values found in the serum [16]. The level of erythromycin lactobionate in the saliva remains in the range of 1.39–4.24 μg/ml when 1 g is administered intravenously every 12 hr for 5 days [17].

Comparative studies on erythromycin and spiramycin done in a mixed population of healthy volunteers and subjects with various oral infections, showed the presence of erythromycin in saliva in an appreciable proportion of the cases (about 65%) one and two hours after administration. A single dose of 500 mg was given orally, and samples were taken after 1, 2, 3, and 6 hr. The highest levels were observed at the end of the first hour (8.37 μg/ml saliva), whereas at the end of the second hour there was a definite drop (5.98 μg/ml saliva) and at the end of the third hour no erythromycin was measurable in most cases. These results indicate that erythromycin appears in saliva very early and in a biologically active form at active levels. The pharmacokinetics of spiramycin are very similar [7, 18]. On the basis of all this it may be concluded that erythromycin shows a particular tropism for the respiratory system and reaches higher levels in the pulmonary tissue than in the serum. Erythromycin also reaches high levels in saliva, but not as high as those seen in pulmonary tissue.

REFERENCES

1. Matsumoto K, Uzuka Y: Concentrations of antibiotics in bronchiolar secretions of the patients with chronic respiratory infections. In: Chemotherapy (Williams JD, Geddes AM, eds.), Plenum Press, New York, London, 1975, 4, p. 73.
2. Fraschini F, Copponi V, Dubini F, Scarpazza G: Concentration of erythromycin and ampicillin in bronchial secretions of patients with chronic respiratory infections. In: Current Chemotherapy (Siegenthaler W, Lüthy R (eds.), American Society for Microbiology Washington DC, 1978, p. 650.
3. May JR, Delves DM: Treatment of chronic bronchitis with ampicillin: some pharmacological observations. Lancet i:929, 1965.
4. Saggers BA, Lawson D: In vivo penetration of antibiotics into sputum in cystic fibrosis. Arch Dis Childh 43: 404, 1968.
5. Saggers BA, Lawson D: Some observations on the penetration of antibiotics through mucus in vitro. J Clin Path 19:313, 1966.
6. Simon C and Clasen I: Sputum concentrations of erythromycin after single and repeated oral administration in adult patients with bronchitis. In: Current Chemotherapy (Siegenthaler W, Lüthy R. (eds.), American Society for Microbiology. Washington DC, 1978, p. 652.
7. Dubini F, Faraone P, Guastamacchia C, Fraschini F: Ricerca della eritromicina e della spiramicina nella saliva. Dental Cadmos 10:1, 1976.
8. Sheila M, Stewart Fisher M, Young JE, Lutz W: Ampicillin levels in sputum, serum and saliva. Thorax 25: 304, 1970.
9. Fraschini F, Braga PC, Copponi V, Scaglione F, Fumagalli G, Gattei G, Scarpazza G: Penetration of erythromycin into the bronchi. Acta Pediatrica Belgica, 1980, in press.
10. Balbirsingh M, Dorn J, Klainer AS, Liss RH, Norman JC, Ward EE: Erythromyrin. In: Current chemotherapy (Siegenthaler W, Lüthy R, (eds.), American Society for Microbiology. Washington DC, 1978, p. 650.
11. Gould JC: Personal communication.
12. Dette G: Personal communication.
13. Canad Med Asso J.: Concentration of orally administered erythromycin and tetracycline in ischemic tissue. Med News in Brief 85:504, 1961.
14. Fraschini F, Braga PC, Copponi V, Gattei G, Guerrasio E, Scaglione F, Villa F, Scarpazza G: Tropism of erythromycin for respiratory system. Communication. In: Current Chemotherapy and Infections Disease. (Nelson JD, Grassi C, eds.), American Society for Microbiology, Washington DC, 1980, p. 659.
15. Fraschini F, Avallon R, Copponi V, Fumagalli G, Mandler F, Scaglione F, Scarpazza G: Bactericidal action of an average dose of erythromycin in the bronchi. Current Med Res Opinion 6:107, 1979.
16. Stephens VC, Puch CT, Davis NE, Hoehn MM, Ralston S, Sparks MC, Thompkins L.: A study of the behavior of propionyl erythromycin in blood by a new chromatographic method. J Antibiot 22:551, 1969.
17. Neaverson MA: Intravenous administration of erythromycin: serum, sputum, and urine levels. Curr Med Res Opinion 4:359, 1976.
18. MacFarlane, JA, Mitchell AAB, Walsh JM, Robertson J: Spiramycin in the prevention of post-operative staphylococcal infection. Lancet i:1, 1969.

DISCUSSION

Dr Butzler: Dr Fraschini, you showed us some very nice slides about the bactericidal effect of erythromycin. But, we have to be very careful in interpreting these results and when we say that erythromycin will be better than, for example, amoxicillin. In Belgium about 10% of H.influenzae have a M.I.C. of more than 12.5 or 25.49, which is a concentration you will never find in the sputum. And we still do not have 10% of β-lactamase producers. Therefore, amoxicillin is still our first choice in treating Haemophilus influenzae infections.

Perhaps an advantage for erythromycin can be noted from your slide. You gave 500 mg erythromycin stearate, three times daily. Our French colleagues apply the dosage of 1 g of erythromycin ethylsuccinate and have marvelous clinical results. We did a study in children and found the same excellent results. When you give 1 g of the erythromycin ethylsuccinate, you have higher concentrations and 2 to 4 g erythromycin ethylsucinate per day is better tolerated than the erythromycin stearate. We have only been able to cure all Haemophilus influenzae infections in children with a dosage of 100 mg per kg per day orally, which is high. I would say that you could give 75 mg per kg per day but not 100.

Would you, in conclusion, agree that erythromycin is, still today, not the first choise in treating Haemophilus influenzae infections?

Dr Fraschini: I agree that erythromycin is not the only antibiotic that can be used in infections of the respiratory tract. But you have to consider that erythromycin is also an antibiotic which is very safe and which has few side effects, in contrast to what has been observed with other antibiotics, particularly the (semisynthetic) penicillins and also the cephalosporins. It is possible to use the new preparation (erythromycin ethylsuccinate) that you have mentioned and to increase the dose. You can obtain very good results, better than those you can obtain with the preparation (erythromycin stearate) I have used in my experiment: I would like to repeat my experiment with this new compound, but it is very difficult to organize.

Dr Stam: How did you account for the blood content of the lung? This may influence your results.

Dr Fraschini: I know this problem. We wash the blood out as much as possible and collect the bronchial secretion from the piece of the lung collected after surgery. The pulmonary tissue we measure was cleaned as much as possible of blood and mucous. An important fact is that the levels of erythromycin in the pulmonary tissue were superior to the levels in the blood, which is different for other antibiotics in which the levels in the tissues are lower than the levels in the blood.

Dr Mouton: Which standards did you use for measuring sputum levels and saliva levels? Did you use a standard of mixed sputum or mixed saliva?

Dr Fraschini: When we measured the antibiotic in the lung, we dissolved it in lung tissue.

Dr Mattie: There have been some reports in the literature that if rat lungs are perfused with erythromycin, the content of erythromycin becomes very high. I think that, in this respect, your data from human results bear out this fact. On the other hand, the animal experiments are more clear-cut because the content of lung tissue can be followed in time. It appears then that a drug like erythromycin–and there are not many of these–is taken up preferably by the lung tissue, probably due to its lipophilia.

I do not know what this implies. If it is taken up preferably by this tissue against a concentration gradient, then this would imply that it is bound to tissue components and the drug that is bound will not contribute to antibacterial efficacy. In this animal experiment the drug was extracted from lung tissue; you used a standard in identical homogenized tissue, so probably what you determined was also total content. Have you any indication of the antibacterial activity of this drug, in the lung tissue? Do you think it is bound inactively to certain binding sites or not?

Dr Fraschini: We wanted to measure erythromycin in pulmonary tissue, because we have found that erythromycin has a bactericidal effect in the bronchi. To justify this bactericidal effect, we wanted to see if erythromycin was concentrated in pulmonary tissue and also in the bronchial secretion. From our results, I can extrapolate that erythromycin is present in an active form, otherwise it is impossible to see this bactericidal effect of erythromycin.

8. PENETRATION OF VARIOUS ANTIBIOTICS INTO SPUTUM

C. SIMON

Various antibiotics that can be administered orally have been investigated in respect of sputum levels by several authors. Table 1 shows that peak sputum concentrations were five times higher after a single dose of 0.2 g minocycline than after the recommended single dose of 0.1 g doxycycline. During continuous treatment with other drugs in the usual dosages, peak sputum levels were highest

Table 1. Sputum levels of various antibiotics.

Drug	Peak concentration in sputum (μg/ml)	Oral dosage (g)	References
Doxycycline	0.33	1 × 0.1	1
Minocycline	2.00	1 × 0.2	2
Erythromycin	0.56	3 × 0.5	3
Cefaclor	0.42	4 × 0.5	4
Cefalexin	0.32	4 × 0.5	5
Amoxycillin	0.50	4 × 0.5	6
Ampicillin	0.20	0.5	7

Table 2. In vitro sensitivity of Haemophilus influenzae to various antibiotics*.

Drug	MIC of Haemophilus influenzae	
	Ampicillin-sensitive strains (μg/ml)	Ampicillin-resistant strains (μg/ml)
Ampicillin	0.5 (0.025–1.0)	32 (16–128)
Erythromycin	1.6 (0.4–3.1)	3.1 (1.6–6.2)
Cefalexin	16 (6.2–32)	16 (6.2–32)
Cefaclor	3.1–6.2	3.1–6.2
Tetracycline	3.1 (0.3)	
Chloramphenicol	0.2–1.5	0.2–1.5

* Mean values, with range between parentheses.

with erythromycin, followed by amoxycillin, cefaclor, cefalexin and ampicillin. In relation to the in vitro sensitivity of Haemophilus influenzae to several antibiotics (Table 2), these findings suggest that the antibiotic concentrations reached are not sufficient to eliminate such bacteria from the sputum in all cases, but we know that host defense mechanisms may enhance the antibacterial efficacy of a drug.

There are many unsolved problems concerning sputum levels in patients with a bronchopulmonary disease. Does doubling the dose produce a corresponding increase of antibiotic concentrations in the sputum? What role does accumulation play during continuous treatment with certain drugs, or is a decrease of antibiotic penetration into the sputum to be expected after improvement of purulent bronchitis? And which pharmaceutical form of a drug, e.g. erythromycin, gives optimal absorption from the intestinal tract and thus guarantees high concentrations in bronchial secretions?

ERYTHROMYCIN

Erythromycin can be administered as stearate or ethylsuccinate. Urine recovery after oral administration is very low (Table 3), i.e., only a tenth of that after intravenous infusion of lactobionate. Comparison of the areas under mean serum level curves after oral and intravenous administration of erythromycin (4 hr $\times$ μg/ml and 24 hr $\times$ μg/ml respectively) showed that only 15% of the oral dose was absorbed from the gut (Fig. 1). Saliva levels run parallel to serum levels, and areas under saliva level curves showed the same correlation between oral and intravenous administration. The most important difference between the stearate and ethylsuccinate forms (Table 4) seems to be the much higher variation of serum levels after administration of the stearate. It is known from the literature and our own studies that 20–30% of the patients given erythromycin as stearate

Table 3. Recovery of various pharmaceutical forms of erythromycin in urine.

Pharmaceutical forms of erythromycin	Route of administration*	Recovery in urine (%)
Stearate	oral	1.4
Ethylsuccinate	oral	1.5
Lactobionate	intravenous	16.0

* dose: 0.5 g.

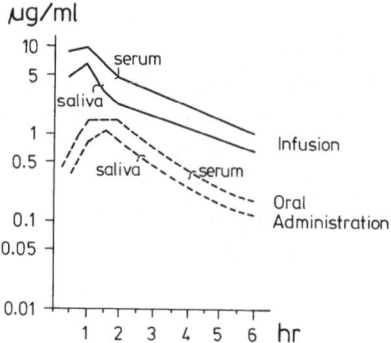

Fig. 1. Mean serum and saliva concentration curves of erythromycin after 1 hr intravenous infusion of 0.5 g of erythromycin lactobionate and after oral administration of 0.5 g of erythromycin ethylsuccinate liquid in six healthy adult volunteers. From Simon and Clasen [3].

are 'non-absorbers', and therefore ethylsuccinate should be preferred for oral administration. Doubling the dose (Table 5) results in a corresponding increase of the area under the serum-level curves (4.1 to 9.1 hr × μg/ml).

To determine sputum levels of erythromycin in 20 patients with purulent bronchitis or pneumonia, volumes of purulent or mucous sputum particles, discarding saliva from the mouth, were mixed with the same volume of a 1% solution of pancreatin. After this mixture had been shaken for 15 min at room

Table 4. Serum levels after oral administration of erythromycin stearate and erythromycin ethylsuccinate in 10 healthy adult volunteers.

Time	Serum levels*	
(hr)	Erythromycin stearate* (μg/ml)	Erythromycin ethylsuccinate** (μg/ml)
0.5	<0.1	0.7 ± 0.04
1.0	0.5 ± 0.29	1.4 ± 0.08
1.5	1.2 ± 0.50	1.5 ± 0.07
2.0	1.2 ± 0.40	1.4 ± 0.07
2.5	1.0 ± 0.30	1.0 ± 0.06
3.0	0.90 ± 0.23	0.70 ± 0.07
4.0	0.77 ± 0.19	0.39 ± 0.05
5.0	0.45 ± 0.11	0.21 ± 0.03
6.0	0.32 ± 0.08	0.13 ± 0.02
Area under curve	(hr × μg/ml) 4.06	(hr × μg/ml) 4.13

* values are means and standard deviation.
** dose: 0.5 g given orally.

Table 5. Mean serum levels after oral administration of erythromycin ethylsuccinate in 10 healthy adult volunteers.

Time	Serum concentration after	
(hr)	0.5 g erythromycin ethylsuccinate (μg/ml)	0.1 g erythromycin ethylsuccinate (μg/ml)
0.5	0.7	1.9
1.0	1.4	2.2
1.5	1.5	2.4
2.0	1.4	2.2
2.5	1.0	2.0
3.0	0.70	1.7
4.0	0.39	1.1
5.0	0.21	0.84
6.0	0.13	0.64
Area under curve	(hr $\times$ μg/mg) 4.13	(hr $\times$ μg/ml) 9.13

temperature, the sputum was completely liquified. After centrifugation for 5 min, the supernatant fluid was investigated, in the same way as serum, by the standard agar diffusion technique with Sarcina lutea (ATCC 9341) and Difco antibiotic assay medium No. 11.

Standard curves with erythromycin base were prepared with pooled liquefied sputum and pooled serum of untreated patients. As shown in Table 6, mean sputum concentrations on the first day of treatment were almost 50% of the

Table 6. Concentration of erythromycin* reached in serum and sputum in 20 adult patients.

Day of treatment	Route of administration**	Level determined in	Mean concentration		
			2 hr (μg/ml)	3 hr (μg/ml)	6hr (μg/ml)
1	oral	serum	0.88	1.19	1.80
		sputum	0.21	0.57	0.56
4	oral	serum	0.61	1.41	1.43
		sputum	0.22	0.41	0.38
1	intravenous	serum	8.70	7.30	2.78
		sputum	1.65	1.27	0.96

 * 0.5 g erythromycin stearate given orally or
** 0.5 g erythromycin lactobionate given intravenously within 1 hr.

mean serum level after 3 hr and about 30% after 6 hr. On the fourth day of treatment, i.e., 2, 3, and 6 hr after oral administration of 0.5 g erythromycin stearate, the mean concentrations in the serum and sputum differed only slightly from the values on the first day of treatment. The ratio between the sputum and serum levels on the fourth day ranged between 25 and 35%. Comparison of patients with mucous and purulent sputum showed no significant difference in antibiotic concentrations in respect of sputum quality.

After a 1-hr intravenous infusion of 0.5 g erythromycin lactobionate (Table 6), sputum concentrations decreased from 1.6 μg/ml (after 2 hr) to 1.2 μg/ml (after 3 hr) and 0.9 μg/ml (after 6 hr). The ratio between the sputum and serum levels changed, as can be expected after intravenous infusion, from 20% (after 2 hr) to 35% (after 6 hr). We have not yet investigated sputum levels after oral administration of erythromycin ethylsuccinate at a dosage of 2 times 1 g daily which is currently recommended for this form in Germany. Information on this point would be of interest in connection with patients with purulent bronchitis, especially when caused by Haemophilus influenzae.

MINOCYCLINE AND DOXYCYCLINE

Similar studies were done with minocycline and doxycycline. After oral administration of 0.2 g minocycline at 24 hr intervals, antibiotic concentrations in purulent sputum were highest after 6 hr (Table 7). On the third day of treatment the 'through-level' (before administration of the next dose) in the serum was 0.32 μg/ml, and the sputum levels after 2 and 3 hr were double those found on the first day. In mucous sputum the concentrations of these antibiotics were much lower,

Table 7. Mean antibiotic concentrations in serum, sputum and saliva after oral administration of minocycline in 19 adult patients*.

Day of treatment	Time	Minocycline concentration in			
		Serum	Purulent sputum	Mucous sputum	Saliva
	(hr)	(μg/ml)	(μg/ml)	(μg/ml)	(μg/ml)
1	2	2.10	0.67	0.07	0.18
	3	2.12	0.97	0.14	0.28
	6	1.69	1.64	0.29	0.21
3	0	0.79	0.32	<0.04	0.17
	2	2.10	1.47	0.10	0.25
	3	2.68	2.02	0.14	—

* Daily dose of 0.2 g minocycline given orally.

and there was no accumulation on the third day. Saliva levels of minocyline were relatively high, which explains the good results obtained in meningococci carriers with this drug for chemoprophylaxis of septicemia and meningitis. After intravenous administration of 0.2 g minocycline in a single dose (Table 8) there was only a small difference in the sputum level compared with oral administration.

With doxycycline, antibiotic concentrations did not differ between purulent and mucous sputum and on the third day were considerably lower than those of minocycline, i.e., only a fourth at 2 hr and a sixth at 3 hr (Table 9). On the third day of treatment the lower dose of 0.1 g led to sputum levels similar to those seen on the first day after 0.2 g. After intravenous injection of 0.2 g doxycycline (Table 10) sputum levels at 2 hr were double those obtained after oral administration of the same dose.

Table 8. Mean antibiotic concentrations in serum, sputum, and saliva after administration of minocycline* in 8 adult patients.

Time	Minocycline concentration in			
	Serum	Purulent sputum	Mucous sputum	Saliva
(hr)	(μg/ml)	(μg/ml)	(μg/ml)	(μg/ml)
1	4.39	0.32	0.11	0.28
2	3.35	0.70	0.24	0.33
3	2.64	0.91	0.31	0.38
6	1.60	1.13	0.22	0.23
25	0.74	0.85	–	–
32	0.55	0.50	–	–

* Single dose of 0.2 g given intravenously within 60 min.

Table 9. Mean antibiotic concentrations in serum, sputum and saliva after oral administration of doxycycline in 18 adults patients.

Dosage daily	Day of treatment	Time	Doxycycline concentration in		
			Serum	Sputum	Saliva
(g)		(hr)	(μg/ml)	(μg/ml)	(μg/ml)
0.2	1	2	3.08 ± 0.28	0.23 ± 0.03	0.39 ± 0.05
0.2	1	3	2.98 ± 0.20	0.27 ± 0.03	0.45 ± 0.05
0.2	1	6	2.48 ± 0.22	0.23 ± 0.04	0.36 ± 0.06
0.1	3	0	1.05 ± 0.16	0.18 ± 0.02	0.16 ± 0.04
0.1	3	2	3.33 ± 0.43	0.33 ± 0.05	0.39 ± 0.07
0.1	3	3	3.19 ± 0.39	0.30 ± 0.04	0.41 ± 0.07

Table 10. Mean antibiotic concentrations in serum, sputum, and saliva after a single intravenous dose* of doxycycline in 10 adults patients.

Time	Doxycycline concentration in		
	Serum	Sputum	Saliva
(hr)	(μg/ml)	(μg/ml)	(μg/ml)
1	5.01 ± 0.40	0.38 ± 0.04	0.50 ± 0.07
2	4.69 ± 0.23	0.42 ± 0.06	0.45 ± 0.07
4	3.62 ± 0.26	0.34 ± 0.05	0.40 ± 0.04

* 0.2 g infused within 3 min.

CEFACLOR

Cefaclor (Panoral, Lilly) is a new oral cephalosporin with stronger in vitro activity than cefalexin for pneumococci, meningococci, E. coli, Klebsiella pneumoniae, and Proteus mirabilis. It also inhibits ampicillin-resistant haemophilus strains at much lower concentrations than cefalexin does. After oral administration of 1 g cefaclor, serum levels rose somewhat faster and became higher than those reached with cephalexin in the first 60 min (Fig. 2). Looking at the end of the second hour, serum concentrations of cephalexin were 50–100% higher. The area under the serum level curve averaged 45 hr × μg/ml for cefaclor and 60

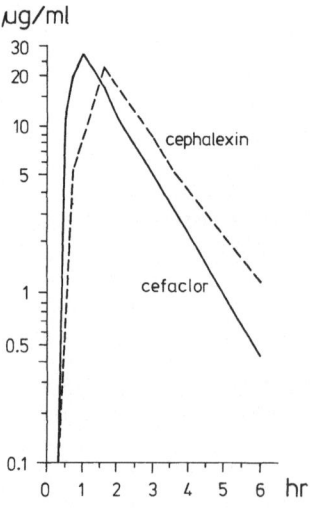

Fig. 2. Mean concentrations of cefaclor (——) and cephalexin (– – –) in the serum of 10 healthy adult volunteers after oral administration of 1 g.

hr × µg/ml for cephalexin. Urine recovery of cefaclor over a period of 9 hr was calculated to be 62%, compared with 85% for cephalexin.

In our study, 15 adult patients with purulent bronchitis or pneumonia received 0.5 g cefaclor four times daily. The serum and sputum concentrations were determined 1, 2, and 3 hr after a single dose on the first day and after repeated administration on the third day. The first dose (0.5 g) resulted in mean serum levels of 8.7 µg/ml after 1 hr, 6.2 µg/ml after 2 hr and 4.5 µg/ml after 3 hr (Table 11). During this period the mean sputum level rose from 0.29 to 0.36 µg/ml. The ratio between the sputum and serum concentrations after 3 hr was 1:12, which was almost the same as the value on the third day of treatment with 4 doses of 0.5 g daily. During repeated administration of 0.5 g or 1 g, peak concentrations of cefaclor in the sputum on the third day were reached earlier (after 2 hr) than after the first dose. This can be explained by a through level reached before administration of the next dose. The higher dose of 1 g cefaclor led to mean peaks of 20.3 µg/ml (first day) and 18.1 µg/ml (third day), and the peaks in the sputum were on average higher than 0.5 µg/ml after the first dose, and this was also the case after repeated administration. Sputum levels in patients were high enough to inhibit the growth of pneumococci, other streptococci, and meningococci from infected bronchi. In vitro experiments performed in our laboratory showed these pathogens to be inhibited by cefaclor levels below 0.4 µg/ml.

CONCLUSIONS

Of the oral antibiotics studied, minocycline and erythromycin penetrate best into purulent sputum, and the relatively high concentrations reached are sufficient for the elimination of sensitive pneumococci. In infections caused by Haemophilus

Table 11. Mean concentrations of cefaclor in serum and sputum after oral administration of 0.5 g (in adult patients) and of 1.0 g (in 10 adults patients) on the first and third day of treatment.

Day of treatment	Dose (g)	Level determined in	Mean concentrations after			Mean individual peaks (µg/ml)
			1 hr (µg/ml)	2 hr (µg/ml)	3 hr (µg/ml)	
1	0.5	Serum	8.7	6.2	4.5	10.6
		Sputum	0.29	0.28	0.36	0.44
3	4 × 0.5	Serum	7.5	8.1	4.3	10.6
		Sputum	0.37	0.42	0.30	0.54
1	1.0	Serum	11.9	13.2	8.5	20.3
		Sputum	0.14	0.27	0.61	0.57
3	4 × 1.0	Serum	11.2	15.2	12.4	18.1
		Sputum	0.23	0.42	0.39	0.54

influenzae clinical failure may occur because the sensitivity of some strains to these antibiotics is too low. After repeated administration, both minocycline and doxycycline accumulate in the sputum. Cefaclor may be more efficient than cefalexin, due to the higher sputum levels reached and the stronger activity against Pneumococcus and Haemophilus influenzae. Amoxycillin is absorbed better than ampicillin and reaches higher concentrations in bronchial secretions than the latter does.

REFERENCES

1. Simon C, Sommerwerck D, Friehoff J: Der Wert von Doxycyclin bei Atemwegsinfektionen (Serum-, Speichel-, Sputum-, Lungen- und Pleuraexsudatspiegel). Praxis und Klinik der Pneumologie 32: 217, 1978
2. Sommerwerck D, Simon C, Friehoff J: Minocyclin zur Therapie von Atemwegsinfektionen. Dtsch Med Wschr 103: 822, 1978.
3. Simon C, Clasen I: Sputum concentrations of erythromycin after single and repeated oral administration in adult patients with bronchitis. Current Chemother 652, 1978.
4. Simon C, Gatzemeier U: Serum and sputum levels of cefaclor. Med J 55: 30, 1979.
5. Halprin GM, McMahon SM: Cephalexin concentrations in sputum during acute respiratory infections. Antimicrobial Agents Chemother 3:703, 1973.
6. May JR: Chemotherapy of Chronic Bronchitis and Allied Disorders. Engl. Universities Press, London, 1968, p. 33.
7. Saggers BA, Lawson D: Some observations on the penetration of antibiotics through mucus in vitro. J Clin Pathol 19:313, 1966.

DISCUSSION

Dr. Mattie: I think the data you showed illustrate what is often encountered in the literature: that concentrations in the sputum already decline when they are nowhere in the vicinity of plasma concentrations. First of all, one would like to know the protein binding in serum, because one may expect that if protein binding does not prevent diffusion at all, it may hinder it. In relation to the non protein-bound fraction, the differences might be much smaller especially for minocycline which is very strongly bound to serum proteins, for the cephalosporins, it would probably not make much of a difference. If even then there would be a parallel decline and no equilibration, do you agree that this could be caused by profuse secretion of bronchial mucus, thus diluting the sputum? If that would be the case, then it would have nothing to do with the drug, but everything with the patient. Therefore my question is, did you collect the sputum quantitatively, or did you just take small samples?

Dr Simon: We took small samples. It was not possible to determine, in our patients, the daily volume of sputum.

Dr Mattie: Could you make an educated guess how much the sputum is diluted by bronchial secretions during the period of sampling?

Dr Simon: We know that dizziness caused by central nervous system distur-third day, and in some patients we did not obtain any sputum on the third day. It was different in each individual.

Dr Kayser: Could you comment on the side effects of minocycline?

Dr Simon: We know that dizziness caused by central nervous system disturbance is possible, but this does not last very long, it may be experienced at the height of the maximum concentration. In bedridden patients, there is no problem but in ambulatory patients, it may constitute a problem.

Dr Sundberg: Dr Mattie raised a very important question: the importance of protein binding. Perhaps I can give you an indirect answer. Acidocillin, a semi-synthetic penicillin is 85% bound to protein, while ampicillin is 15% bound to serum proteins. There are no important differences in the penetration characteristics of these two drugs. So the degree of protein binding does not seem to have any major effect on the penetration.

Dr Mattie: You cannot say that just because two different drugs have the same degree of penetration, that protein binding does not make a difference. There are many other drugs that are not protein bound at all and nevertheless show differences in penetration. One way to prove your point could be for instance, to change the protein binding of one of the drugs by means of competition by a second drug.

Your example may just as well imply, that the more proteinbound drug must have a much better penetration, because the protein-bound fraction can not penetrate.

Dr Sundberg: With regard to acidocillin and ampicillin no differences, as far as I know, have been found. Even in secretions, the same degree of penetration takes place, as far as I know.

Dr Simon: May I answer you in this connection? We compared doxycycline and minocycline, which had nearly the same high degree of protein binding, but, because minocycline is more lipophilic and penetrates better, the sputum levels were higher than in doxycycline.

Dr Van der Waaij: My question concerns the absorption of the two oral cephalosporins: cephalexin and cefaclor. I can imagine that you could have a recovery of 60% with cefaclor and 85% with cephalexin, due to the fact that cefaclor is less stable or metabolized in the body. Then, you will find a difference in recovery in the urine. Could that be the explanation?

Dr Simon: We examined stability of cefaclor in serum and urine. If serum is kept cool or frozen, the stability is very good. This is not the case with urine, where the PH changes; the recovery may be somewhat higher. We also determined the urine concentrations by high-pressure liquid chromatography and found a good correlation. I think cefaclor is metabolized in the body to some degree and, therefore, the urine recovery is also lower for this reason.

Dr Van der Waaij: Is the difference in recovery not a difference in absorption, rather than a difference in the fate of the molecule–the metabolism–inside the body?

Dr Thompson: Is there any comparative study that shows that a drug with a better penetration into the sputum has superior clinical results?

Dr Simon: There are clinical studies comparing the efficacy of two drugs, for instance, amoxycillin and erythromycin. I think there is a relatively good agreement between the relatively low antibiotic concentrations in sputum and the failure in cure in distinct infections.

Dr Mattie: There is a point which is often brought up, namely as the patient gets better, the constitution of the sputum changes, and antibiotic concentrations become lower with less purulence. Again, this might be explained by several causes, but also by technical problems, for instance. One cannot always make out from the context what the case may be. Do you always use sputum as a standard for determination?

Dr Simon: There is a difficulty in determining sputum levels in the course of the treatment because the sputum alters; it is easier to have exact values on the first day. In minocycline, we found a lower concentration on the third or fourth day, and then we used mucous sputum for preparing our standards.

9. PENETRATION OF VARIOUS ANTIBIOTICS INTO THE MIDDLE EAR

L. SUNDBERG and S. ERNSTSON

INTRODUCTION

In spite of the enormous body of literature on different aspects of otitis media, there have been surprisingly few studies dealing with the important problems of antibiotic penetration into the middle ear. As late as 1966, when Silverstein et al. [1] published their article in Pediatrics, they could begin with the words: 'To date, there has been no proof that antibiotics reach the middle ear in effective concentrations during the course of active inflammation in that area'. This pioneer work resulted in further research on the penetration of different antibiotics in acute otitis media and secretory otitis media [2–4]. However, most investigations deal with the penetration in the input phase, and often in a small number of cases. In 1979, a larger study was published, which described characteristics of the penetration of an antibiotic through the middle-ear mucosa, not only in the first period after administration of the drug (input phase) but also in two subsequent periods, the steady-state phase and the output phase [5].

AETIOLOGY

According to Paparella [6], acute otitis media can be classified after the nature of the middle-ear effusion in purulent otitis media, serous otitis media, and mucoid or secretory otitis media.

It should be emphasized that these different types of otitis media have in common an inflammatory aetiology. In the normal healing process one type often changes gradually to another before total resolution takes place.

The number of epidemiological reports on the aetiology of acute otitis media is overwhelming [7–11] (Table 1). It must now be considered as an established fact that the usual relevant pathogens are pneumococci, Haemophilus influenzae, β haemolytic streptococci group A, and Branhamella (Neisseria) catarrhalis. The pathogenity of B. catarrhalis is no longer a controversial question [7, 12]. Staphylococcus aureus in samples of middle-ear effusions should be regarded as a contamination from the ear canal. In exceptional cases, cultures of middle-ear

Table 1. Literature on the aetiology of acute otitis media published between 1964 and 1980.

Author	Year	Number of cases	Pneumo- cocci	H. influenzae	Strepto- cocci	Mixed pathogens
			(%)	(%)	(%)	(%)
Grönroos et al. [27]	1964	381	31.8	11.0	14.4	2.9
Grönroos et al. [27]	1964	153	23.5	26.8	8.5	6.8
Degré et al. [28]	1965	1312	30.5	11.3	7.5	3.3
Coffey et al. [12]	1966	267	34.5	27	1.9	8.9
Norstedt [29]	1967	278	42	16	3	
Lundgren et al. [18]	1967	324	31	11	2.5	
Nylén et al. [30]	1969	237	24	32	4	16
Kamme et al. [7]	1971	71	50.7	14.7	5.3	
Lundgren [10]	1972	660	36	13	3	
Bergholtz et al. [31]	1972	450	29	11	2.4	28
Frölund et al. [8]	1975	147	19.9	22.6	8.2	
Nylén [9]	1975	320	28.8	30	4.7	
Brook [32]	1979	168	36.9	30.1	5.5	
Haugsten et al. [33]	1980	297	33	17	7	
Total		5065	32	16	6	5

exudates yield gram-negative bacilli, and in infants aged up to three months coliforme bacteria are a usual finding.

In many studies on the frequency of the various pathogens, pneumococci have been isolated from the middle-ear effusions in 30–60%, H. influenzae in 15–20%, B. catarrhalis in about 10%, and group A streptococci in about 5% [13]. B. catarrhalis is often found as the causative agent during the first three years of life. It has been argued that H. influenzae is the most common pathogen in children under two to five years, but recent studies have shown that pneumococci are the dominant aetiological agent at all ages.

ANTIBIOTICS AND THE MINIMUM INHIBITORY CONCENTRATION OF BACTERIA IN VITRO

In the light of the aetiological data it is rather natural that betalactam antibiotics and macrolides have been used as first-line drugs in the treatment of acute otitis media. Among the betalactam antibiotics, the penicillin G and V, the semisynthetic azidocillin, and broad-spectrum penicillins such as ampicillin and amoxicillin, are the most commonly used [26]. Erythromycin is in this context the leading [15] macrolide. Other groups of antibiotics have not gained popularity.

The minimum inhibitory concentration in vitro (M.I.C.) is different in different strains of the same bacterium.

The gram-positive cocci are very sensitive to penicillins and macrolides, and are generally inhibited by in vitro very low concentrations of these antibiotics. Naturally-occurring, penicillin-resistant streptococci have not been described [11]. In one study on 33 strains of pneumococci, the M.I.C. value for penicillins V and G, azidocillin, and ampicillin ranged from 0.0008 to 0.125 mg/l, and in 35 strains of group A streptococci the M.I.C. value for the same antibiotics between 0.008 and 0.25 mg/l [14]. The extreme sensitivity of pneumococci and group A streptococci to penicillin was confirmed by another study, where the M.I.C. values were 0.01 to 0.04 and 0.001 to 0.05 mg/l, respectively [11]. Erythromycin, too, is higly effective in vitro against pneumococci and group A streptococci. The M.I.C. value for pneumococci ranges between 0.001 and 0.04 mg/l and for streptococci between 0.008 and 0.25 mg/l [15]. B. catarrhalis is not as sensitive to penicillin V as the gram-positive cocci. The M.I.C. value of penicillin V ranges between 0.15 and 4.8 mg/l, but 80% of 108 strains were inhibited by 0.3 mg/l of penicillin V [11].

On the other hand, significantly higher antibiotic levels are required for the inhibition of H. influenzae. This gram-negative rod often possesses a pronounced resistance to both penicillin V and erythromycin. Ampicillin and amoxicillin have, however, a remarkably good antibacterial activity against this pathogen. Thus, in 399 strains of H. influenzae, with or without capsule, the M.I.C. value for ampicillin was as low as 0.032 mg/l, whereas the corresponding M.I.C. for azidocillin and penicillin G ranged between was 0.064 and 64 mg/l and for penicillin V between 0.25 and 128 mg/l [14]. In another study, 80 strains of H. influenzae isolated from middle-ear exudates of children with acute otitis media, were incubated in 8–10% CO_2, to reproduce the environment thought to exist in the middle ear. Penicillin V in a concentration of 9.6 mg/l inhibited 95% of the strains [11]. The M.I.C. value in that investigation ranged between 1.2 and 19.2 mg/l.

A relatively new and serious problem is the betalactamase-producing strains of H. influenzae, which are totally resistant to all penicillins including ampicillin and amoxicillin. However, a combination of ampicillin and clavulanic acid seems to have a certain effect against these strains. As a betalactamase-resistant drug, erythromycin could be expected to have an antibacterial effect on betalactamase-producing strains of H. influenzae, but this is still a moot question. In general, rather high concentrations of erythromycin are needed to inhibit strains of H. influenzae. Thus, the reported M.I.C. values range between 0.5 and 25.0 mg/l in one study and in another between 0.19 and 3.12 mg/l [15].

PENETRATION OF PENICILLINS

In 1966, Silverstein et al. were the first to demonstrate the penetration of antibiotics into middle-ear effusions [1]. Using an agar cup method which was not totally exact because of a difficult dilution procedure, they showed detectable amounts of antibiotics in effusions from patients with otitis media. All but one of 14 patients with acute purulent otitis media had received a single intramuscular injection of 500 000 IU of penicillin G (the other patient received 1 000 000 IU). The concentration of penicillin in the middle-ear effusions one to three hours later appeared to be equal to or higher than the M.I.C. for the majority of strains of streptococci and pneumococci, but not that for H. influenzae.

Six other patients with suppurative otitis media received a single oral dose of either penicillin V or penicillin G, and the penicillin concentration in the effusions was determined between two and a half to eight hours later. The concentration of penicillin in the effusions surpassed the M.I.C. for the majority of strains of streptococci and pneumococci in three of the patients, but did not exceed that for H. influenzae in any of the patients.

A single intramuscular dose of 500 000 IU penicillin G was given to each of four patients with secretory otitis media. Three of these four patients showed detectable amounts of penicillin in the effusions one to two hours later, in two cases even in the range of the levels found in patients with purulent otitis media. In one case of secretory otitis media penicillin V was administered by the oral route and antibiotic activity was found in the effusion after about two hours. The same authors also studied the penetration of oxytetracycline. In twelve patients with purulent otitis media the concentration of oxytetracycline in the middle-ear effusion one to three hours after an intramuscular injection of 100 mg was equal to or higher than the M.I.C. of the majority of strains of streptococci in only six patients, for pneumococci this occurred in ten patients, and for H. influenzae in none.

Four other patients with secretory otitis media were given 100 mg oxytetracycline in a single intramuscular injection, and a just barely detectable antibiotic activity, too low for quantitive determination, was found in the effusions one and a half to three hours later.

The interpretation of this study was brilliant, and the conclusions have had fundamental importance in the field of research on middle-ear penetration. The authors established that antibiotic penetration is promoted by the presence of an active inflammation in the mucosa, such as occurs at the onset of acute otitis media. In secretory otitis media the less active inflammation makes the penetration slower. These observations have been confirmed by later studies [16, 17]. Thus, penicillin and oxytetracycline penetrated into middle-ear effusions in patients with acute otitis media considerably better than in those with secretory

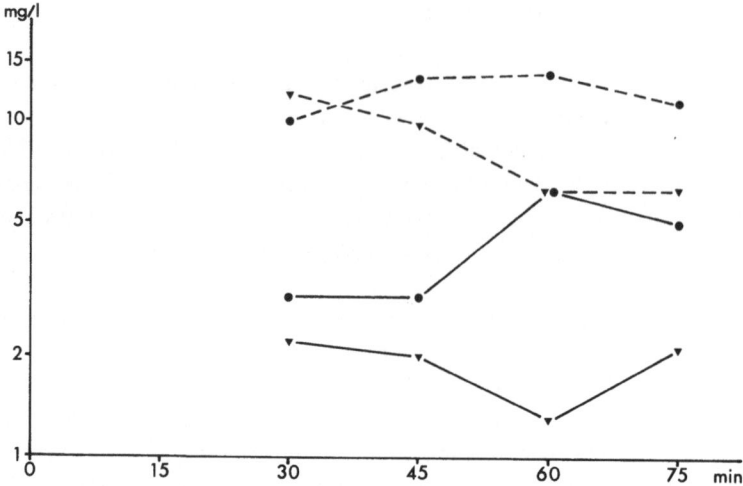

Fig. 1. Concentration of penicillin V in serum and in middle-ear in patients with acute otitis media [3]. Dosage 13 mg/kg: concentration in serum ▲—▲ and in effusions ▲ – ▲. Dosage 26 mg/kg: concentration in serum ●—● and in effusions ● – ●.

otitis media. The penicillin levels in the middle-ear effusions in purulent otitis media exceeded the M.I.C. values much further than the corresponding values for oxytetracycline did, and consequently penicillin should and has always been preferred to oxytetracycline for the management of acute otitis media.

Silverstein et al. also put into focus the very important problems presented by the often very high M.I.C. values of many strains of H. influenzae to conventional drugs. This problem should remain in the centre of interest for many years to come. In concluding, these authors formulated a kind of research programme with the following words: 'Other antibiotics should be studied to find an agent that both enters the middle ear in concentrations equal to or in excess of those needed to inhibit H. influenzae in vitro and possesses the activity of penicillin against gram-positive pathogens responsible for middle-ear infections'.

One method used in the attempts to raise the antibiotic levels in middle-ear effusions above the M.I.C. of most strains of H. influenzae was to apply a higher dose of penicillin V. As early as 1966, Silverstein et al. [1] suggested that 'penicillin, perhaps in doses somewhat larger than those usually employed against the highly sensitive gram-positive cocci, may be effective in the management of many, if not most, infections due to H. influenzae'. In a clinical study in 1967, Lundgren and Rundcrantz showed that the healing rate in otitis media increased significantly when a larger dose of penicillin V was employed [18]. In 1969, Kamme et al. [3] confirmed these observations and demonstrated a dose-response relationship: a higher dose of penicillin V gave a higher concentration in the middle-ear effusion (Fig. 1). In 54 children with acute otitis media, 38 middle-ear effusions and 54 serum samples were assayed by an agar well method. After a single oral dose of 13 mg per kg body weight penicillin V (20 000 IU), the maximum of the mean concentrations in the middle-ear effusions was 2.1 mg/l and was reached after 30 min simultanously with the peak in the serum. After a single, oral dose of 26 mg per kg body weight (40 000 IU) penicillin V, the peak in the effusions had a mean value of 6.3 mg/l at 60 min. The peak serum level occurred at the same time. Two hours after administration, the antibiotic levels in effusions and in serum were closely similar. The concentration in effusion was three times higher after the larger (doubled) dose. This amount (6.3 mg/l) of penicillin V in the effusions should be enough to inhibit not only the gram-positive cocci but also about 90% of strains of H. influenzae, according to Kamme [11].

In 1970, Lahikainen published his interesting results on the penetration of penicillin in patients with acute otitis media, secretory otitis media and chronic otitis media [16]. The material consisted of 206 middle-ear samples from acute otitis media, 22 middle-ear samples from secretory otitis media, and 45 middle-ear samples from chronic otitis media. All these patients received a single intra-muscular injection of 400 000 IU penicillin G and, at intervals ranging from half an hour to twelve hours, the amount of antibiotic in the effusions was assayed with an agar cup method. Blood samples were collected at the same time.

In acute otitis media penicillin penetrated readily due to the active inflammation and hyperaemia in the middle-ear mucosa. The maxima occurred simultanously in the serum and effusion after one hour. The average value in the effusions was 1.27 units/ml with a range of 8.2 to 0 units/ml, the corresponding concentration in serum being 3.58 units/ml with a range of 11.2 to 1.2 units/ml. After twelve hours, the mean value in the effusions had decreased to 0.17 units/ml and in the serum to 0.03 units/ml. These findings indicate that antibiotic levels are maintained longer in the effusions than in the blood and that elimination of the drug is slower from the effusion than from the blood. This is in full agreement with other observations [5]. In secretory otitis media, Lahikainen found even slower penetration. After one hour the effusions of only 3 of the 13 patients showed traces of penicillin and after two to four hours 6 out of 9 patients. This slower penetration is explained by the absence of active inflammation in the middle-ear mucosa. In chronic otitis media penicillin was found in only 28 of 45 effusions. The low incidence of positive samples was explained by the presence of betalactamase-producing bacteria inactivating penicillin. Thus, penicillin did not seem to be a drug of choice in the management of infections in chronic otitis media.

The changes occurring in the middle-ear mucosa during the course of acute otitis media affect the penetration of the antibiotic. Initially, penetration is good because the presence of bacteria sustains the active inflammation accompanied

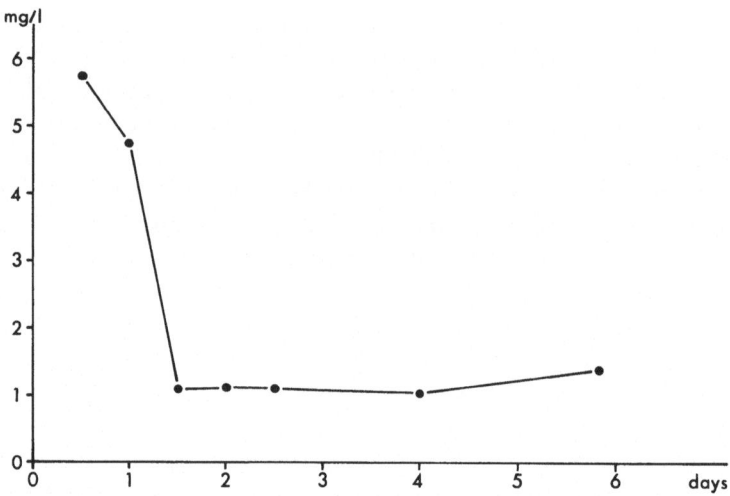

Fig. 2. Penicillin V in middle-ear effusion during the course of acute otitis media [19].

by capillary dilatation, oedema, and exudation in the mucosa. Later in the course, there is a considerable decrease of the penetration when the phase of active inflammation subsides and the mucous membrane becomes more like the mucosa in secretory otitis media, showing an increase of the number of goblet cells and mucus production.

Lundgren and Rundcrantz [19] studied the penetration of penicillin V throughout the course of acute otitis media (Fig. 2). The drug was given to 17 patients in a dose of 26 mg per kg body weight every twelve hours. The initial mean peak value in the effusions was reached after about one day and was 5.7 mg/l. However, after two and a half days of treatment the mean concentration in the effusions had decreased to 1.4 mg/l despite the steadily high antibiotic level of 12.9 mg/l in the serum. These findings underscore the importance of knowing the exact phase of an acute otitis media in which antibiotic levels are measured in a penetration study.

THE PENETRATION OF AZIDOCILLIN, AMPICILLIN AND AMOXICILLIN

Azidocillin is a semisynthetic penicillin considered to be more effective then penicillin V against H. influenzae. In a study covering more than 500 strains of H. influenzae, Forsgren found that 90% of the strains were inhibited by 0.63 mg/ml and that the corresponding value for penicillin V was 5.6 mg/ml [20]. In 1973, Lahikainen reported on the penetration of azidocillin into the middle ear in cases of acute otitis media. One hour after a single standard oral dose of azidocillin, the

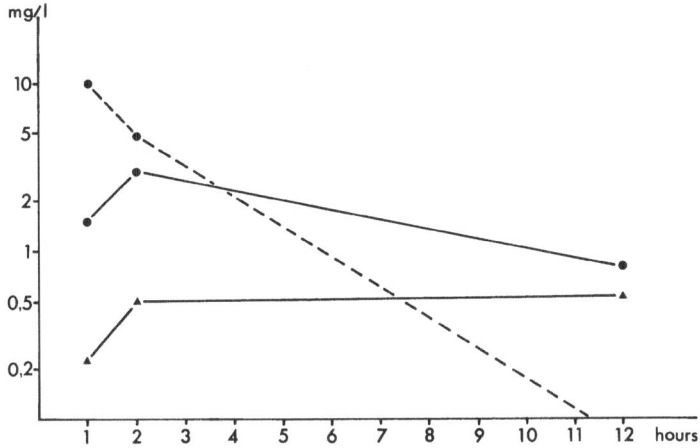

Fig. 3. Concentration of azidocillin in serum (●—●) and in effusions of patients with acute otitis media (●—●) and secretory otitis media (▲ — ▲) [17].

antibiotic level in the effusions was higher than the M.I.C. value of almost 50% of the strains of H. influenzae, ranging between 0 and 3.4 mg/l [21]. In 1977, Lahikainen et al. [22] published a more comprehensive study on the penetration of azidocillin into the middle ear (see below).

Ampicillin soon attracted much interest because of its documented good activity against H. influenzae. All of 80 strains of H. influenzae were inhibited by a concentration of 0.60 mg/l, and 97% of the strains by 0.30 mg/l. The corresponding amounts of penicillin V inhibited no strains at all [23]. Coffey studied the penetration of ampicillin in ten children with acute otitis media [2]. After a single intramuscular dose of 250 or 500 mg, depending on the body weight, the effusion concentrations after 60–80 min ranged between 1.6 and 19.0 mg/l. These levels were well above the M.I.C. values of the bacteria usual causing acute otitis media, including H. influenzae.

Lahikainen et al. reported in 1977 on the penetration of ampicillin and azidocillin [17]. A single oral dose of either 15 mg azidocillin or 10 mg ampicillin per kg body weight was administered to 101 patient with acute otitis media and 63 with secretory otitis media. For the acute otitis media effusions the mean concentration of azidocillin (Fig. 3) was 1.56 mg/l after one hour and 3.21 mg/l after two hours. The mean concentration of ampicillin (Fig. 4) was 1.15 mg/l after one hour and 2.17 mg/l after two hours. For the effusions in secretory otitis media the mean concentration of azidocillin after one hour was 0.22 mg/l and after two hours 0.50 mg/l. The corresponding mean concentrations of ampicillin were 0.17 mg/l and 0.23 mg/l, respectively.

These findings confirmed earlier observations that the penetration of antibiotics through the middle ear mucosa is better in acute otitis media than in

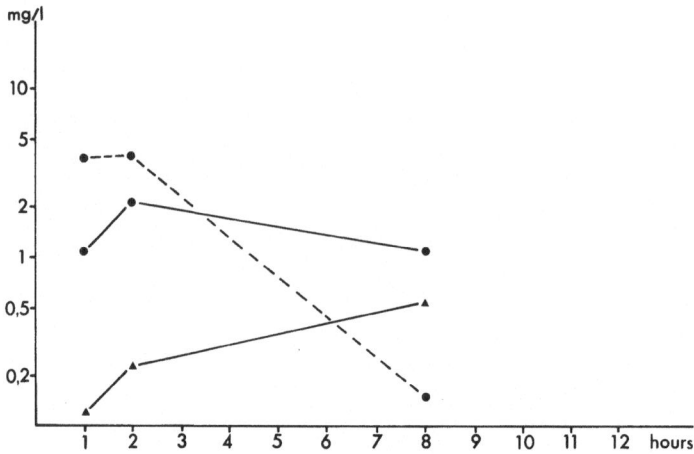

Fig. 4. Concentration of ampicillin in serum (●—●) and in effusions of patients with acute otitis media (● – ●) and in secretory otitis media (▲ – ▲) [17].

secretory otitis media and that antibiotics persist longer in effusions than in the blood. The amounts of antibiotic in the effusions were higher then the M.I.C. of most strains causing otitis media except in secretory otitis media, where the level was below the M.I.C. of most strains of H. influenzae. Finally, the penetration of ampicillin and azidocillin was similar, although only 15% of ampicillin is bound to proteins compared to 85% for azidocillin.

In 1977, Klimek et al. compared the penetration of ampicillin with that of amoxicillin, a newer broad-spectrum penicillin with about the same antibacterial activity as ampicillin [24]. Amoxicillin is, however, absorbed better than ampicillin and reaches higher serum levels. A single oral dose of 1000 mg of either amoxicillin or ampicillin was administered to 28 children with secretory otitis media. The concentration in the effusions one to two hours later was significantly higher for amoxicillin than for ampicillin, the mean value being 6.20 mg/l for amoxicillin and 1.43 mg/l for ampicillin. It is possible that the increased diffusion gradient of amoxicillin contributes to this better penetration.

PENETRATION OF ERYTHROMYCIN

In 1971, Bass et al. reported on the penetration of various esthers of erythromycin in acute otitis media [25]. Four children received erythromycin ethylsuccinate and 4 were given erythromycin estolate, both in an oral dose of 12.5 mg per kg (i. e., 50 mg/kg/day) for 24 h at six-hour intervals. Two hours after the fourth (last) dose the serum and effusion concentrations were determined and a bacteriological assay of the middle-ear effusion was performed at the same time. In the 4 children treated with erythromycin ethylsuccinate the level in the serum ranged between 0.45 and 2.60 mg/l with a mean of 1.30 mg/l, and that in the effusions ranged between 0.24 and 1.02 mg/l with a mean of 0.84 mg/l. In all four cases gram-stained bacteria were found and cultures yielded growth of H. influenzae in one case and of B. catarrhalis in another. In the children treated with erythromycin estolate the serum levels ranged between 4.14 and 12.33 mg/l with a mean of 7.55 mg/l; the effusions showed considerably higher levels ranging between 1.68 and 8.0 mg/l with a mean of 4.18 mg/l. In the patient with the highest amount of erythromycin in both serum and effusion, culture yielded growth of H. influenzae; in the three other patients no bacteria were cultured. The better penetration of erythromycin estolate can to a certain extent be explained by the fact that the peak of the erythromycin estolate ester occurred after two hours, i.e., the time when the samples were drawn, whereas the corresponding peak of the erythromycin ethylsuccinate ester occurred after one hour.

The levels of both erythromycin esters in the effusions were well above the M.I.C. of pneumococci and group A streptococci, and the cultures did not yield gram-positive cocci, but cultures of three of the eight children yielded gram-negative bacteria: H. influenzae in two and B. catarrhalis in one. Thus, as in treatment with penicillin V, the problem of eradicating all strains of H. influenzae in the middle-ear effusions has not yet been solved.

These findings are in agreement with the clinical observations made by Howie and Ploussard [26]. In otitis media treated with erythromycin the authors obtained excellent results where the pathogens were gram-positive cocci. In otitis media due to H. influenzae, however, persistant strains were found in the effusions. The question as to whether the penetration was insufficient or the strains were resistant was not answered by that investigation. In 1979, Sundberg et al. published a large series comprising 108 cases of secretory otitis media in which the pharmacokinetics of erythromycin ethylsuccinate were studied in the input phase, in the steady state phase and in the output phase [5] (Fig. 5). Erythromycin ethylsuccinate was administered orally three times a day for various periods in the recommended standard dosage. The concentration of the drug was determined in effusions and blood with an agar well diffusion method.

In the early input phase there was a slow penetration of erythromycin into the middle-ear effusion. Two hours after the initial dose the mean value was less than 0.13 mg/l. However, twelve hours later, i.e., two hours after the second dose, the mean level was 0.6 mg/l. Not until the late input phase, after 26 and 38 h, did the concentration in the effusion reach its plateau level of about 1.1 mg/l. This concentration was equal to the mean plasma peak level. This level was maintained in the middle ear during the steady state: 1.2 mg/l on the second day and 1.1 mg/l on the tenth day. The output phase was characterized by a slow elimination. Thus, 14 h after the last dose of a ten-day course, the mean middle-ear effusion level was still as high as 0.9 mg/l, indicating that a steady state had

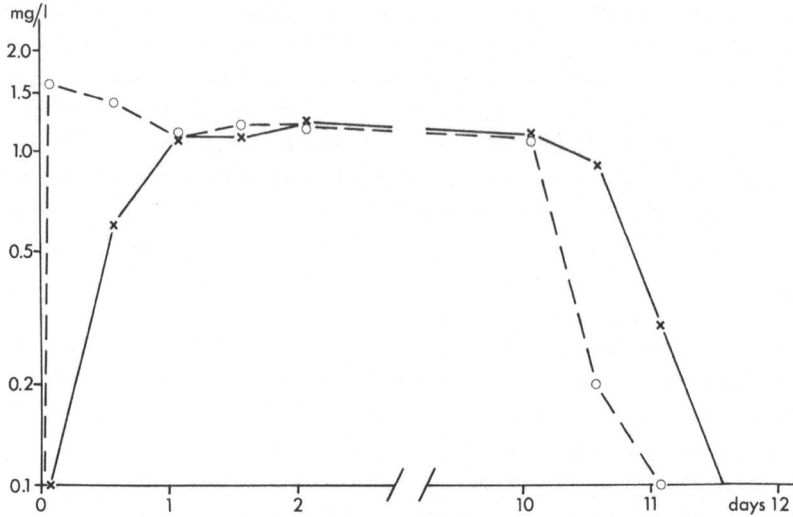

Fig. 5. Erythromycin concentration in middle-ear effusions (x – x) and in plasma (O—O). All values are means of ten or more determinations except on day 11½, which concerns only two cases.

been established in the middle ear. However, 26 h after the last dose, elimination had continued and the concentration of erythromycin had decreased to 0.3 mg/l and after another twelve hours there was no detectable antibiotic activity in the effusion.

These findings are in agreement with other observations [16, 17] concerning the pharmacokinetics of antibiotics. It has been established, that the penetration of drugs into middle-ear effusions is considerably slower in secretory otitis media than in acute otitis media and that the concentrations are maintained much longer in the effusions than the blood.

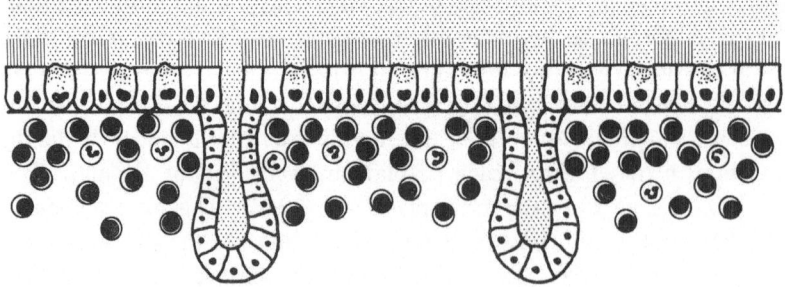

Fig. 6. Schematic representation of respiratory-tract mucosa in secretory otitis media, chronic sinusitis, and chronic bronchitis. Shaded area denotes mucous blanket [5].

The concept that the respiratory mucosa reacts uniformly to an inflammatory stimulus has also been put forward (Fig. 6). The same uniform mode of response is seen in secretory otitis media, in sinusitis and bronchitis. Therefore, secretory otitis media can be regarded as a suitable model for the study of antibiotic penetration into respiratory mucosa, and the results can be expected to be valid for inflammatory diseases in any area lined with respiratory epithelium.

CONCLUSION

The aim of successful antibiotic therapy in acute otitis media is to achieve antibiotic concentrations in the middle ear mucosa and effusion that lie higher than the M.I.C. values of the relevant bacteria. Knowledge of the antibiotic concentration in the focus of inflammation as well as the M.I.C. value of the aetiological agent is of basic importance for the management of acute otitis media.

All of the available investigations show a wide range of the values obtained, the lowest and highest usually differing by a factor of ten, due partially to technical errors. There is also a wide individual divergence in the pharmacokinetics of antibiotics, and the dosage must be determined with great care. Even then, there will be some individuals who do not conform to the general pattern and fail to attain the expected antibiotic levels.

It should be mentioned that in 75–80% of all cases of acute otitis media recovery occurs spontanously without treatment, that 75–80% of the cases of acute otitis media are caused by pathogens sensitive to penicillins and macrolides, and the remaining 15–20% due to H. influenzae can be treated successfully with amoxicillin. But there are clouds on the horizon. An increasing number of reports on betalactamase-producing strains of H. influenzae suggest that a serious problem may be impending.

REFERENCES

1. Silverstein H, Bernstein JM, Lerner PI: Antibiotic concentrations in middle ear effusions. Pediatrics 38:33, 1966.
2. Coffey Jr JD: Concentration of ampicillin in exudate from acute otitis media. Pediatrics 72:693, 1968.
3. Kamme C, Lundgren K, Rundcrantz H: The concentration of penicillin in serum and middle ear exudate in acute otitis media in children. Scand J Infect Dis 1:77, 1969.
4. Sundberg L. Edén T, Ernstson S: Penetration of erythromycin into middle ear secretions. A preliminary report. Curr Med Res Opin 5, Suppl 2:28, 1978.
5. Sundberg L. Edén T, Ernstson S, Pahlitzsch R: Penetration of erythromycin through respiratory mucosa. A study using secretory otitis media as a model. Acta Otolaryngol Suppl, 365, 1979.

110

6. Paparella MM, Middle ear effusions: definitions and terminology. Ann Otol Rhinol Laryngol 85, Suppl 25:8, 1976.

7. Kamme C, Lundgren K, Mårdh P-A: The acetiology of acute otitis media in children. Scand J Dis 3:1, 1971.

8. Frölund Thomsen V, Sederberg Olsen J, Sörensen H, Thomsen J: Bacteriology and antibiotics in acute suppurative otitis media. J Otolaryngol 5:289, 1976.

9. Nylén O: Otitis media acuta. En klinisk, bakteriologisk och serologisk studie, Gotab, Kungälv, Sweden, 1975.

10. Lundgren K: Otitis media acuta hos barn. En klinisk studie av etiologi och terapi. Studentlitteratur, Lund, Sweden, 1972.

11. Kamme C: The aetiology of acute otitis media in children and the relationship between bacterial characters and the clinical course in penicillin V therapy. Staffanstorps Tryckeri AB, Sweden, 1971.

12. Coffey Jr JD, Martin AD, Booth HN: Neisseria catarrhalis in exudate otitis media. Arch Otolaryngol 86:69, 1967.

13. Kamme C: Bakteriologiska synpunkter på akuta och recidiverande otiter. In: Otiter och otosalpingiter. Ett ÖNH-symposium i Lund, September 1977. Bröderna Ekstrands Tryckeri AB, Lund, Sweden, 1978, p. 167.

14. Tunevall G: Jämförelse mellan aktiviteten hos olika penicilliner och cefalosporiner, Symposium om diagnostik och behandling av akut otit, sinuit och tonsillit. Stockholm 1973. Bröderna Ekstrands Tryckeri AB, Lund, Sweden, 1974, p. 109.

15. Dette GA: Vergleich der Gewebegängigkeit von Erythromycin. Infection 7 (3):129, 1979.

16. Lahikainen EA: Penicillin concentration in middle ear secretion in otitis. Acta Otolaryngol 70:385, 1970.

17. Lahikainen EA, Vuori M, Virtanen S: Azidocillin and ampicillin concentrations in middle ear effusion. Acta Otolaryngol 84:227, 1977.

18. Lundgren K, Rundcrantz H: Otitis media acuta hos barn. Läkartidningen 64:63, 1967.

19. Lundgren K, Rundcrantz H: Microbiology in serous otitis media. Ann Otol Rhinol Laryngol Suppl 25, 85:152, 1976.

20. Forsgren U: Penicillinkänslighet hos Haemophilus. Symposium om riktad och kontrollerad terapi med penicilliner, Lidingö 1969. Astra Läkemedel AB, Södertälje, Sweden, 1969.

21. Lahikainen EA: Azidocillinkoncentration i mellanöresekret. Symposium om diagnostik och behandling av akut otit, sinuit och tonsillit. Stockholm 1973. Bröderna Ekstrands Tryckeri AB, Lund, Sweden, 1974, p. 209.

22. Kamme C: Susceptibility in vitro of Haemophilus influenzae to penicillin G, penicillin V and ampicillin. Incubation of strains from acute otitis media in air and in CO_2-atmosphere. Acta Path Microbiol Scand 75:611, 1969.

23. Klimek JJ, Nightingale C, Lehmann WB, Quintilani R: Comparison of concentrations of amoxicillin and ampicillin in serum and middle ear fluid of children with chronic otitis media. J Infect Dis 135(6):999, 1977.

24. Bass JW, Steele RW, Wiebe RA, Dierdoff EP: Eythromycin concentrations in middle ear exudates. Pediatrics 48 (3):417, 1971.

25. Howie VM, Ploussard JH: The 'in vitro sensitivity test'–bacteriology of middle ear exudate. Pediatrics 44:940, 1969.

26. Ekedahl C, Holm S, Holmberg K, Rundcrantz H: Hur bör penicillin ges vid otit, sinuit, tonsillit? Läkartidningen 77(16):1520, 1980.

27. Grönroos J, Kortekangas A, Ojala L, Vuori M: The aetiology of acute middle-ear infection. Acta Otolaryngol (Stockh) 58:149, 1964.

28. Degré M, Ulstrup J: Bakteriefunn ved otitis media. T norske Laegeforen 85:994, 1965.

29. Norstedt S: Otitis media acuta hos barn i allmänpraktik. Läkartidningen 64:787, 1967.

30. Nylén O, Branefors-Helander P, Herberts G: Värdering av antibiotisk effekt vid otiter och sinuiter. Symposium om riktad och kontrollerad terapi med penicilliner. Lidingö, 1969. Astra Läkemedel AB, Södertälje, 1969.

31. Bergholtz L, Rudberg R: Behandling av otitis media acuta. Läkartidningen 69:3922, 1972.

32. Brook I: Otitis media in children: a prospective study of aerobic and anaerobic bacteriology. Laryngoscope 89:992, 1979.

33. Haugsten P, Lorentzen P: The bacterial etiology of acute suppurative otitis media. J Laryngol Otol 94:169, 1980.

DISCUSSION

Dr Grote: I want to ask Dr Sundberg first whether he has used the model of middle-ear effusion just to study the penetration of erythromycin, or whether he is stating that, especially in cases of secretory otitis media, which is different from acute otitis media, antibiotic treatment is the first treatment of choice. I think, particularly in the case of Haemophilus influenzae, which is difficult to attack, that the basis of secretory otitis media is really de-aeration. Do you agree that when we adopt the surgical approach, to re-establish aeration of the middle ear, you do not need antibiotic therapy. Perhaps this is even true for most of the cases of acute otitis media?

Dr Sundberg: Of course, you are right. We are not treating secretory otitis media with antibiotics, but we are using secretory otitis media as a model for the study of antibiotic penetration. We have the same opinion regarding the treatment of both acute otitis media and secretory otitis media, as you have. Secretory otitis media is a very usual disease, and a spin-off effect when you evacuate the middle ear is to use the secretion in penetration studies.

Dr Grote: May I come back again on Haemophilus influenzae, because it gives me the impression that it is one of the very important bacteria in the course of some secretory otitis media cases. The most recent studies establish that it is especially the underaeration which gives the so-called symbionts the possibility to cause secretory otitis media. What is Dr Sundberg's opinion about it?

Dr Sundberg: As I said in my lecture, 15 to 20% of the cases of acute otitis media were due to Haemophilus influenzae, and 30 to 60% are caused by pneumococci. Because of the high M.I.C. of Haemophilus influenzae, this pathogen often is the problem. You have to use other antibiotics like amoxicillin, in cases of otitis media due to Haemophilus influenzae. Research in this field has been focused on how to overcome the M.I.C. of Haemophilus influenzae. That is the real problem; pneumococci and streptococci, are no problem in this respect.

Dr Maclaren: I am a little surprised that you chose penicillin V for Haemophilus influenzae instead of ampicillin or amoxycillin. A long time ago, it has been pointed out that in general, penicillin V and penicillin C were roughly equally active in vitro, that penicillin V is a little less active than penicillin G in Haemophilus influenzae.

Dr Sundberg: That is right. Penicillin V is given orally; it is much easier to administer. Penicillin G must be given by injection. That is the problem. It has shown that if you use a high dose of penicillin V, you get levels in the middle effusions well above the M.I.C. of most strains of Haemophilus influenzae. It is a dose response relationship.

Dr Butzler: We would never give penicillin V to treat Haemophilus infections, because, in vivo, it does not work very well.

Furthermore, it is very dangerous to tell practitioners that the treatment of otitis is either amoxicillin, erythromycin or another drug. I think it is very important to look at the age of the children. In Belgium, we would never have given erythromycin to children between 0 and 2 years; we give it to older children.

Dr Sundberg: I said that 75% of all acute otitis media heal spontaneously and about 75% of the cases are caused by micro-organisms very sensitive both to penicillin V and erythromycin. The problem is the 15–20% of cases due to Haemophilus influenzae, where I recommended amoxicillin.

As to the age group, the impression formerly was that Haemophilus influenzae was the dominating etiological agent during the five first years of life. However, new studies have shown that pneumococci dominate as the etiological agent in all ages. So, then, penicillin V is a very good agent as well as erythromycin.

Dr Van der Meer: What is the evidence that you need to exceed the M.I.C. in the middle ear?

Dr Sundberg: Why do we have bacteriological laboratories?

Dr Van der Meer: That was not my question.

Chairman: Can you give a more definite answer?

Dr Sundberg: The answer is very simple. If you have two drugs, one drug surpasses the M.I.C. in vitro for a special bacterium and the other drug does not, then it is logical to use the drug which surpasses the M.I.C. of a specific pathogen.

Chairman: Does Dr Mattie want to comment on that?

Dr Mattie: Yes, very briefly. Of course, the drug that is the most effective in vitro is probably the most effective in vivo. But, this is only a difference in effectiveness, it has nothing to do with the mystical barrier that has to be surpassed and which is called the M.I.C.. Concentrations below the M.I.C. can very well be efficient.

Another point I want to make is that at least this example shows that, if one waits long enough, equilibration between plasma and any tissue level will be found–at least if tissue fluid is sampled. In this respect, I am not sure whether there is much binding of erythromycin to plasma proteins; if there is, it would be important to know the protein content of the middle-ear effusion in secretory otitis media, because it would be very peculiar if there would be a high degree of plasma protein binding and a sort of accumulation in the middle ear.

Another question pertaining to erythromycin: would it be possible that the form in which you administer the drug–an inactive ester, for instance, which is often done–would penetrate into the middle ear as an inactive ester, but very lipophilic, and after that be hydrolized and as such give high concentrations, being trapped as it were it the middle ear?

Dr Sundberg: I do not know, but I think there is hydrolysis before the penetration into the middle ear. It is possible that there is some kind of trapping of erythromycin in the middle ear, but we cannot prove it in our studies at all.

Protein binding is a problem: You can discuss it as long as you want to. However, all the time, there is an equilibrium between the part which is free and the part which is protein-bound. I think this is the main thing in this connection.

Dr Mattie: I think we should keep another thing in mind. Serum concentrations are determined some time after sampling. If hydrolysis of the inactive ester is slow, then it will go on between the sampling time and the time of the essay. This is a well-known problem which, however, is often overlooked in the case of erythromycin. This would give a false impression of the level of active erythromycin at the moment of sampling.

Dr Ernston: Erythromycin in the middle ear fluid was determined in a bacteriological assay. Furthermore, we tested to see if something happened between the time it was taken to until 14 days afterwards, while being stored in a refrigerator. No change in antibiotic level was found at all. So, at least under the conditions that we had, there was no ongoing hydrolysis. Blood samples and the middle-ear samples were taken within one minute of each other. The middle-ear fluids were taken around the peak of the serum level, that is about two hours after the last given dose.

Dr Mattie: That was not the point that I wanted to make. The time it takes to determine the level is long enough to get complete hydrolysis. The only thing is to make sure that, at the moment you draw the sample, hydrolysis is stopped first by blocking hydrolyzing enzymes, but as far as I know this is never done.

Chairman: Thank you for this addition.

Dr Kerrebijn: There is a question which preoccupies me all the time when we speak about tissue penetration, and that is that we cannot easily obtain tissues. My question is: could white blood cells be used as a model for tissue and is it possible to measure concentration of antibiotics in the white blood cells, or do antibiotics not penetrate through the white blood cell membrane? I know nothing about this, but it must be possible, when they penetrate, to measure concentrations because some of the techniques we have at present available should make this not too difficult.

Dr Mattie: May I answer this question? You only mentioned leukocytes as a model for tissue, the site of the infections we are speaking of now, is not inside the cell; tissue is not identical with cells. Tissue infections are infections of the extracellular space, so in this respect leukocytes would not be a good model.

Dr Grote: I would like to come back to the pharmacodynamics in the model of secretory otitis media. We have different types of secretory otitis, different types of middle-ear effusion, which have different protein contents. Which type did you use: the secretory type?

Dr Sundberg: Most of the cases had a middle-ear effusion which was very tenacious, very mucoid, but there were some cases too which had a more serious secretion in the middle-ear, so we had both types of secretion in our study. There were no differences, as far as we could see.

10. PENETRATION OF VARIOUS ANTIBIOTICS INTO SINUS CAVITIES

O. KALM

INTRODUCTION

Initially, it was thought that orally or intramuscularly administered antibiotics did not reach the sinus mucosa in therapeutic concentrations [1]. In 1968, however, Lundberg and co-workers described a series of sinusitis patients treated with doxycycline whose sinus secretions showed appreciably higher concentrations of the antibiotic than the minimum inhibitory concentration (M.I.C.) for the isolated bacteria [2]. Since then, a number of mainly Scandinavian publications have reported penetration of many other antibiotics in therapeutic concentration into sinus secretions and mucosal tissue. All of the published reports concern investigations done in secretions and mucosal tissue of maxillary sinuses. For practical reasons, the present discussion will be limited to the penetration properties of penicillins, tetracyclines, erythromycin, and ampicillins, the antibiotics used most frequently in the treatment of upper respiratory infections.

BACTERIOLOGY

Bacteriological investigations in acute and chronic maxillary sinusitis have shown pathogenic bacteria in the secretions of 60–90% of the cases. The latter percentage is the result of a wider use of anaerobic culture techniques in recent years. The pathogens found most frequently in both acute and chronic sinusitis are Streptococcus pneumoniae and Haemophilus influenzae, which are about equally common and have been isolated in 50–60% of the cases studied. Group A beta-haemolytic streptococci–formerly a common finding–are at present isolated in not more than 1–10% of the cases [3–5]. Where the use of anaerobic culture techniques has increased, anaerobic bacteria have been isolated alone or together with aerobic pathogens in 10–33% [4,5]. Anaerobes have been reported to be the only likely pathogen in up to 25% of the cases [4].

PENETRATION STUDIES

Concentrations of antibiotics in sinus secretions and mucosal tissues are usually measured in vitro with a paper-disc or agar-well diffusion method. The investigated antibiotics were administered orally or intramuscularly, and various derivatives of the same antibiotic were tested.

Penicillin
It has been firmly established that the M.I.C. values of penicillin V for group A streptococci and pneumococci are 0.02 and 0.06 μg/ml, respectively, but 3.0 μg/ml is needed to inhibit 80% of the strains of Haemophilus influenzae. About 4% of all Haemophilus influenzae strains now produce betalactamase and thus are resistant to all of the usual penicillin derivatives.

In the first report concerning orally administered penicillin, i.e., phenoxymethylpenicillin acid given in a dose of 1.6 g every 24 hr, measurable concentrations in sinus secretions were found in only 10 out of 17 patients after 2–3 days of treatment. The specimens were taken 1–3 hr after the last dose was given [6]. Lundberg and co-workers sought a practical method for comparison of the antibacterial effect of different antibiotics in the sinuses. Since not only the observed concentrations but also the M.I.C. values of the antibiotics had to be taken into consideration, they chose a unit defined as 'adequate' concentration of the antibiotic based on an international study of antibiotic sensitivity testing, sponsored by WHO, where the bacterial sensitivity was classified into four groups (1–4 in order of decreasing sensitivity). Group 1 comprises bacteria with a high degree of sensitivity for the antibiotic used, making in vivo response probable when mild to moderately severe systemic infections were treated (usually via the oral route) [7]. For example, in group 1, the limit for 'adequate' concentration of penicillin is 0.25 μg/ml and for tetracyclines 1.0 μg/ml. For these two antibiotics, group 1 includes most of the upper respiratory pathogens (pneumococci, streptococci, Haemophilus influenzae). These investigations showed 'inadequate' concentrations of penicillin in a high percentage of sinus secretions after a single dose or repeated doses of penicillin V, benzylpenicillin, and procaine benzylpenicillin, occurring at various intervals between administration and the last sampling [8,9]. The concentrations in the sinus mucosal tissue seem to be more reliable, giving adequate concentrations in about 80% of the samples. The concentrations in the blood almost always exceeded those in the secretions [10]. In a kinetic study, Ekedahl et al. [11] determined the concentrations of penicillin V in the maxillary sinus mucosal tissue 45, 60, and 90 min. after oral administration of 25 mg/kg body weight. Although their series was too small to be conclusive, all patients showed a clear tendency for the antibiotic level to stabilize above 2 μg/g at 90 min. The concentration of active antibiotic was thus considerably higher than the M.I.C. for most of the bacteria responsible for the sinusitis. Lower concentrations occurred more often in infected than in uninfected mucosa.

Tetracyclines

The in vitro M.I.C. values of doxycycline (the most commonly used tetracycline derivative) for group A streptococci and pneumococci are about 0.5 μg/ml and 0.25 μg/ml, respectively. Of the various strains of Haemophilus influenzae, 80% are inhibited by 2.0 μg/ml [12].

According to the first report on the subject [2], doxycycline administered in a dose of 100 mg twice daily appeared in secretions of all 12 patients in a concentration ten times higher than the M.I.C. of the bacterium in question. For tetracyclines, the concentration limit defined as 'adequate' for bacteria in group 1 is 1.0 μg/ml [7]. In studies on tetracyclines given alone or together with penicillin in single or repeated doses, 'adequate' concentrations occurred significantly more often with tetracyclines, and with the combination more often in mucous than in purulent secretions (Table 1). The combined therapy consisted of penicillin V 0.4 g four times a day and tetracycline HCl 0.25 g four times a day [9]. Measurable concentrations of both antibiotics were present in the secretions for a least 6 to 8 hr [8]. In relation to the administered dose tetracyclines, and especially doxycycline, always give higher and more reliable concentrations in the secretions than penicillin does [2,9,13]. In the mucosal tissue adequate concentrations of penicillin are found in 80% of the secretions. The mucosa-to-serum ratio of tetracyclines is significantly higher than that of penicillin, and the concentration in the mucosa always seems to exceed that in the secretions [10].

Erythromycin Stearate

The penetration of erythromycin stearate into maxillary sinus secretions or mucosa has not been investigated often but, unlike that of penicillin, led to measurable concentrations in all cases [12,14,15]. An oral dose of 500 mg twice daily for 2 to 4 days gives a mean concentration of 0.6 μg/ml in the secretion. At a dose of 500 mg three times a day, the concentration in secretions and mucosa doubles, but the serum concentration does not change, which indicates a ten-

Table 1. Comparison of 'adequate' concentrations in purulent and mucous sinus secretions*.

	Number of secretions	
	Purulent	Mucous
Penicillin	2/18	10/15
Tetracycline	16/18	15/15

* From Lundberg and Malmborg, 1973 [9].

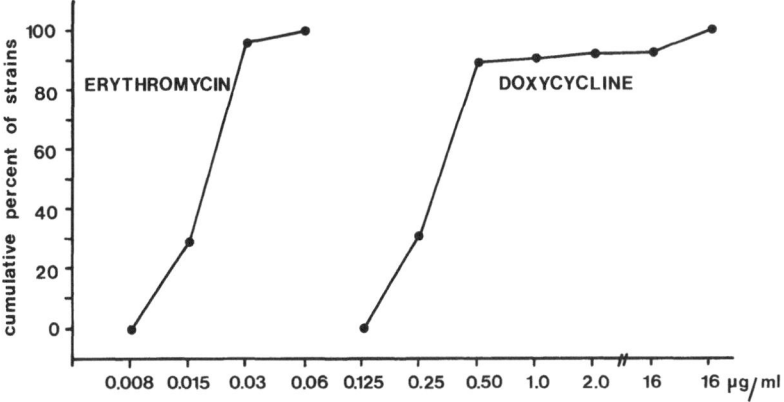

Fig. 1. Sensitivity (M.I.C.) of group A streptococci (100 strains) to erythromycin and doxycycline.

dency for this antibiotic to accumulate in the sinus when a high dose is given (Table 2). At a relatively high dose, i.e. 500 mg three times a day, the mean concentration in mucosal tissue amounts to 1.8 μg/g [14]. One author has found much higher concentrations in the mucosa even after single dose [15]. Comparison of the sensitivity of group A streptococci (Fig.1), group A pneumococci (Fig.2), and H. influenzae (Fig.3), to erythromycin stearate and doxycycline, done for 100 strains of each species, showed that compared with doxycycline, erythromycin is far more effective than doxycycline against group A streptococci and pneumococci but less effective against H.influenzae. At a concentration of 1 μg/ml, erythromycin inhibited 32% of the strains and doxycycline 70% [12]. However, the frequency of tetracyclineresistant group A streptococci and pneumococci in Sweden has been reported to be about 10% and 1–3%, respectively [16].

Table 2. Concentrations of orally administered erythromycin stearate in serum, sinus secretions, and mucosal tissue of the sinus.*, **

Dose	Number of patients	Time after last dose (hr)	Concentration in serum (μg/ml)	Concentration in secretion (μg/ml) and mucosa (μg/g)
2 × 500 mg	10	4.4	2.2	0.6
3 × 500 mg	10	4.7	2.2	1.3
3 × 500 mg	15	2.5	2.3	1.8

* From Kalm et al., 1975 [12] and Paavolainen et al., 1977 [14].
** MIC values for the most frequently found bacteria – see Figs. 1, 2, 3.

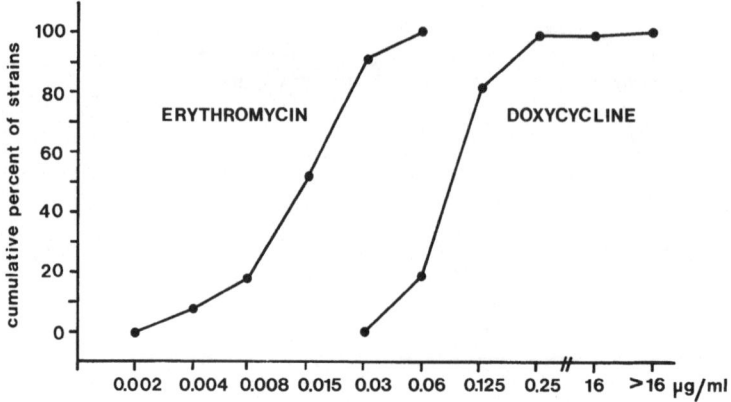

Fig. 2. Sensitivity (M.I.C.) of pneumococci (100 strains) to erythromycin and doxycycline.

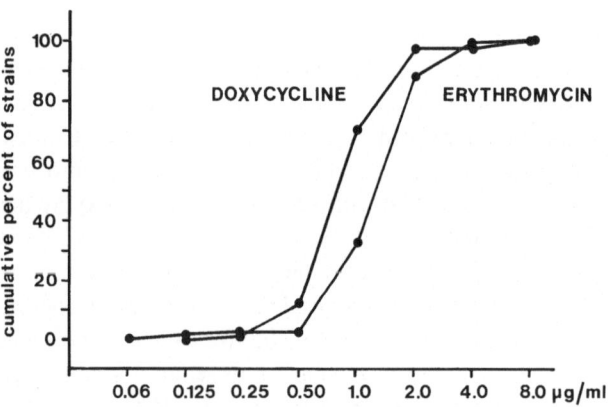

Fig. 3. Sensitivity (M.I.C.) of Haemophilus influenzae (100 strains) to erythromycin and doxycycline.

AMPICILLINS

It is well known that in vitro the M.I.C.s for group A streptococci and pneumococci are 0.03–0.06 μg/ml and about 1.0 μg/ml for H. influenzae. About 4% of the strains of H. influenzae produce betalactamase and hence are resistant.

Penetration of this group of antibiotics into sinus cavities has been studied by only a few investigators in small series. Four patients given ampicillin 0.5 g four times daily showed very low concentrations in secretions [17]. In another study, four days of treatment with an oral dose of 0.5 g three times daily gave secretion levels as high as 1.49 to 2.98 μg/ml, depending on whether the material was obtained by irrigation or aspiration [18]. For the mucosa, dose-dependent concentrations between 0.63 and 0.74 μg/ml have been reported [19,20]. Bacampicillin gives mean peak serum concentrations about three times higher than ampicillin does and the ratio between the concentration in mucosal tissue and the peak serum concentration seems to be roughly 1:3 [11,21].

DISCUSSION

The results of in vitro studies on antibiotic penetration into sinus mucosal tissues and secretion can be summarized as follows.

In relation to dose the tetracyclines, especially doxycycline, give the highest concentrations in both secretion and mucosal tissue and a favorable serum-to-mucosa ratio of about 1.0 [9, 22].

Penicillins have a lower level of penetration [9,10], but penicillin V in high single doses (25 mg/kg body weight) gives concentrations well over the M.I.C., at least in the mucosa, for most of the commonly isolated bacteria [11].

Erythromycin stearate, used in conventional doses, penetrates well into both secretions and mucosa, but only in the mucosal tissues it gives concentrations well over the M.I.C. for most strains of H. influenzae [12,14,15].

For ampicillins, especially bacampicillin, it is probable that the penetration level gives concentrations well over the M.I.C. for most strains of bacteria isolated.

The differences between the ability of antibiotics to penetrate have been partially explained as depending on differences in lipid solubility. Doxycycline, for example, has a much higher lipid solubility than penicillin and is consequently better able to penetrate poorly vascularized tissue [8].

Since the desired and specific effect of antibiotic treatment is the elimination of bacteria from the infected tissue, some authors have attempted to determine whether there is any real correlation between the concentration of antibiotics in the sinuses and their capacity to eliminate the bacteria in the secretions. During treatment, Carenfelt et al., found a marked reduction of the number of viable bacteria in secretions when the antibiotic concentrations were higher than the

Table 3. Elimination of bacteria from sinus secretions in relation to 'adequate' concentrations of certain antibiotics.*

	Penicillin		Tetracycline		Doxycycline	
	<0.25	⩾0.25	<1.0	⩾ 1.0	<1.0	⩾1.0
Antibiotic concentration (μg/ml)						
No. of sinus secretions studied	7	8	2	10	2	8
No. of sinus secretions with viable bacteria	6	4	2	4	2	3

* From Eneroth et al., 1975 [22].

above mentioned adequate concentrations (Table 3). It must be kept in mind, however, that in these investigations the concentrations exceeded the M.I.C. values of the isolated bacteria by a factor of ten or more [23]. Bacteria have sometimes been found in the sinus secretions even though the antibiotic concentration was much higher than the M.I.C. of the strain in question [6,12,13]. The concentrations of penicillin and tetracycline have also been reported to be lower in purulent than in mucous secretions [9,10]. The reason for these somewhat contradictory results is not entirely clear, but in all probability many complex mechanisms are involved. For example, studies on the local gas composition in sinusitis have shown pO_2 values close to zero and high pCO_2 values in purulent sinus secretions, and these findings were associated with acid pH and heavy growth a facultative anaerobes such as pneumococci. Such conditions even interfere with the local protective function of the mucosa and with the bactericidal function of granulocytes [23,24]. Locally produced protective factors such as IgA are also seen in smaller amounts under such conditions, probably as a result of the higher concentration of proteolytic enzymes in purulent secretions compared with mucous secretions [25].

In frequently diagnosed diseases, often treated with antibiotics such as sinusitis, all clinicians should give preference to an effective antibiotic with as narrow an antibacterial spectrum as possible. This is important from an ecological point of view, to avoid an unnecessary augmentation of the number of antibiotic-resistant bacteria.

REFERENCES

1. Strong MS, Tonkin RW: The treatment of maxillary sinusitis by local injection of procaine penicillin-in-oil. J. Laryngol Otol 65:809, 1951.
2. Lundberg C, Gullers K, Malmborg AS: Antibiotics in sinus secretions. Lancet 2:107, 1968.
3. Kortekangas AE: Antibiotics in the treatment of maxillary sinusitis. Acta Otolaryngol. Suppl 188:379, 1964.
4. Van Cauwenberge P, Verschraegen G, Van Renterghem L: Bacteriological findings in sinusitis (1963–1975). Scand J Infect Dis Suppl 9:72, 1976.
5. Carenfelt C, Lundberg C, Nord CE, Wretlind B: Bacteriology of maxillary sinusitis in relation to quality of the retained secretion. Acta Otolaryngol 86:298, 1978.
6. Gullers K, Lundberg C, Malmborg AS: Penicillin in paranasal sinus secretions. Chemotherapy 14:303, 1969.
7. Ericsson H, Sherris JC: Antibiotic sensitivity testing. Report of an international collaborative study. Acta Pathol Microbiol Scand (B). Suppl 217:74, 1971
8. Lundberg C, Malmborg AS: Concentration of penicillin and tetracycline in maxillary sinus secretion after a single dose. Scand J Infect Dis 6:79, 1974.
9. Lundberg C, Malmborg AS: Concentration of penicillin V and tetracycline in maxillary sinus secretion after repeated doses. Scand J Infect Dis 5:123, 1973.
10. Lundberg C, Malmborg AS, Ivemark BI: Antibiotic concentrations in relation to structural changes in maxillary sinus mucosa following intramuscular or peroral treatment. Scand J Infect Dis 6:187, 1974.

11. Ekedahl C, Holm SE, Bergholm AM: Penetration of antibiotics into the normal and diseased maxillary sinus mucosa. Scand J Infect Dis Suppl 14:279, 1978.

12. Kalm O, Kamme C, Bergström B, Löfkvist T, Norman O: Erythromycin stearate in acute maxillary sinusitis. Scand J Infect Dis 7:209, 1975.

13. Axelsson A, Brorson JE: Concentration of antibiotics in sinus secretions. Doxycycline and spiramycin. Ann Otol 82:44, 1973.

14. Paavolainen M, Kohonen A, Palva T, Renkonen OV: Penetration of erythromycin stearate into maxillary sinus mucosa and secretion in chronic maxillary sinusitis. Acta Otolaryngol 84:292, 1977.

15. Kaminszcick I, Galan H: Interés del estudio de los niveles tisulares de antibióticos en sinusopatias. Semana Med 138:1280, 1971.

16. Kahlmeter G, Kamme C: Tetracycline-resistant group A streptococci and pneumococci. Scand J Infect Dis 4:193, 1972.

17. Gnarpe H, Lundberg C: L-phase organisms in maxillary sinus secretions. Scand J Infect Dis 3:257, 1971.

18. Axelsson A, Brorson JE: The concentration of antibiotics in sinus secretions. Ampicillin, cephradine and erythromycinestolate. Ann Otol 83:323, 1974.

19. Jeppesen F, Illum P: Concentration of ampicillin in antral mucosa following administration of ampicillin sodium and pivampicillin. Acta Otolaryngol 73:428, 1972.

20. Burckhardt F, Thumfart W, Waller G: Untersuchungen zum Übergang von Antibiotika (Doxycyclin, Ampicillin) in die polypös-hyperplastisch veränderte Kieferhöhlenschleimhaut. Laryng Rhinol 58:347, 1979.

21. Giebel W, Schönleber KH, Breuninger H, Ullman U: A comparison of the pharmacokinetics in serum and nasal secretions after oral bacampicillin and ampicillin. Scand J Infect Dis Suppl 14:285, 1978.

22. Eneroth CM, Lundberg C, Wretlind B: Antibiotic concentrations in maxillary sinus secretions and in the sinus mucosa. Chemotherapy 21, Suppl 1:1, 1975.

23. Carenfelt C, Lundberg C: Aspects of the treatment of maxillary sinusitis. Sc and J Infect Dis Suppl 9:78, 1976.

24. Carenfelt C, Lundberg C: Purulent and non-purulent maxillary sinus secretions with respect to pO_2, pCO_2 and pH. Acta Otolaryngol 84:138, 1977.

25. Carenfelt C, Lundberg C, Karlén K: Immunoglobulins in maxillary sinus secretion. Acta Otolaryngol 82:123, 1976.

DISCUSSION

Chairman: Dr Stam would like to present a short survey of his studies on the pharmacokinetics of antibiotics in the pleural cavity.

Dr Stam: In collaboration with Professor D.M. Maclaren, our group studied the penetration of antibiotics into pleural fluid. Pleural fluid is continuously formed and absorbed. The mean hydrostatic pressure in the capillaries of the parietal pleura, which derive from the aortic circulation, is higher than the mean hydrostatic pressure in the capillaries of the visceral pleura, which derive from the pulmonary and bronchial arteries. This pressure difference between the capillaries of these two pleural layers and the negative intrapleural pressure are responsible for the production of pleural fluid. The osmotic pressure of the plasma exceeds the hydrostatic pressure in the capillaries of the visceral pleura. The pleural fluid produced is therefore reabsorbed (Fig. 1).

In volunteers with a pleural effusion due to a malignancy we estimated the serum and pleural levels of ampicillin after two doses of bacampicillin (400 mg and 800 mg) given orally. Individual differences between the results were remarkable but consistent. The peak level in pleural fluid was lower and delayed compared with the serum level. However, when the peak level is reached the

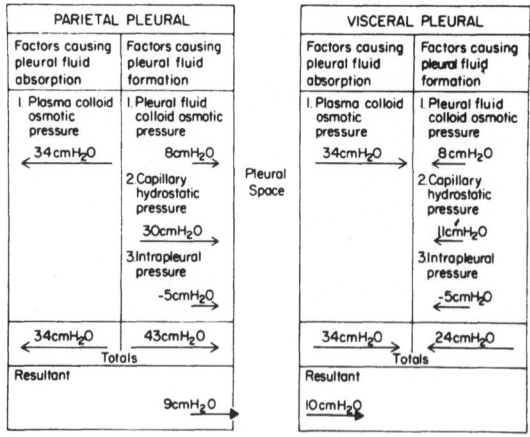

Fig. 1. Physiological properties of the pleura.

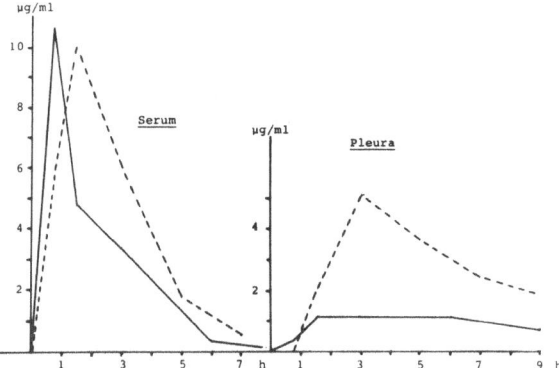

Fig. 2. Serum and pleural fluid concentrations after oral administration of 400 (———) and 800 mg (————) bacampicillin.

concentration remains stable, given a plateau (Fig. 2). Even 12 hr after a single oral dose of 800 mg bacampicillin the pleural fluid level lies well above the M.I.C. of H. influenzae. In our opinion, pleural fluid is comparable with tissue fluid, and we recommend its use for the study of penetration of antibiotics.

Chairman: Thank you. I would like to ask what are the arguments that the pleural cavity is a good model for measuring tissue concentration?

Dr Stam: It is just the same. Also in the tissues you have capillaries with high pressure, the fluid is coming out of the arterial side of the capillary and flows back again at the venous side.

Chairman: I wonder what Dr Mattie thinks about this?

Dr Mattie: Yes, I did point out earlier to Dr Stam that there is the same misunderstanding that can be found in the literature on tissue cages. It is not that they would not contain tissue fluid, but the compartment is not a good model for the tissue compartment, because its volume is much larger. Therefore the pleural cavity is only a good model for the pleural cavity.

Of course, by studying the effusion of the drug into the pleural cavity, some pharmacokinetic properties of the drug might be deduced, but these concentrations do not reflect the concentrations in tissue spaces. There is a relationship between the dimensions of the tissue space that is studied and the delay in the course of the concentrations. One can imagine that the delay for the drug to reach the pleural effusion from the blood is much greater, where you have a bigger volume, than when it has to diffuse into the ordinary tissue spaces. The distance between capillaries in a tonsil or in a lung or wherever, is of course much and much smaller–only a few micrometers–than the distance between the site where the drug comes into the pleural effusion and the site where it goes out, apart from the question that the pleural fluid is not being mixed; it is stationary and the drug has to diffuse over a large area. One should keep in mind that diffusion over a relatively small area into a relatively large volume is quite a disadvantage. This is for instance why the lung has so many alveoli; just to make diffusion over a large area possible.

Chairman: Thank you. Now I would like to continue the discussion and focus on Dr Kalm's paper.

Dr Ernston: I would like to say that this is really an example of sinus secretion measurements. It is an example where pharmacokinetics pay off in clinical practise. In Sweden we treated sinusitis during the '50s and '60s with penicillin, nosedrops and irrigation the first, third and fifth day and so on, until the sinuses were clear. If the sinus did not clear up after five irrigations, the patient was considered a chronic case and was operated on. Then Dr Lundberg of Karolinska, Stockholm, started using tetracycline because he was curious as to whether this drug did anything at all in cases of acute sinusitis. And it did not, but tetracycline penetrates in quite another way. So, after a few months, we stopped and instead treated cases of acute sinusitis with one irrigation first, to settle the diagnosis, then we did not repeat the irrigating and just waited until the course was completed ten days later. We saw the patients fourteen days after the initial irrigation, and most of them were cured. Later, we found not only that tetracycline was miraculously effective in cases of acute sinusitis, but also erythromycin. We believe now that this is mainly because of its lipophylic characteristics, and the ability of tetracyclines and erythromycin to penetrate even later in the course of an active inflammation, as has been shown by investigations in secretory otitis media and in the sinusitis cases, too. So, this is really an example where pharmacokinetics has paid off in better treatment of patients.

Dr Kalm: I think part of this discussion will come up tomorrow, after the discussion on the treatment of sinusitis. There are very many etiological viewpoints on this disease. We have talked much about bronchitis. The maxillary sinus is different in many ways. It has been shown, for example, that in purulent maxillary sinusitis the oxygen tension is about 0 and the carbondioxide tension is very high. There, you will have an acid pH and promotion of a growth of facultative anaerobes such as pneumococci and Haemophilus influenzae. Lower concentrations have been shown of immunoglobulins, especially IgA and the complement C3 and C4, which seem to be necessary for opsonization. So, the function of the granulocytes is also impaired in this situation. Perhaps that is part of the explanation as to why we can find bacteria in secretions where the concentration of the antibiotic is 30 to 50 times higher than the M.I.C..

Dr Sundberg: I will just use the figures of Dr Kalm as an indirect proof that our model works. The concentration of erythromycin in sinus secretions, which Dr Kalm achieved, is very close to our figures which we had as concentration in the middle ear. We think the reason is because it is the same respiratory mucosa in the middle ear as in the sinus.

Dr Ernston: May I comment on this? Because of the properties of respiratory mucosa, instead of using the fluid from the middle ear, we use the adenoid tissue. We find the same concentration for several days during erythromycin therapy. We get the same figure in the sinuses and inside middle ear spaces. So at different sites the respiratory mucosa behaves in a similar way, at least as far as erythromycin is concerned.

Dr Grote: May I comment on Dr Sundberg's remarks about the similar mucosa of the middle ear and the sinus? I do not agree with him. As he knows, there is a complete difference between the mucus of the middle ear and the mucus of the sinus, depending on what type of secretory otitis media and what type of metaplasia you have. If you have the mucus type, as you have investigated, you will find ciliated mucosa, but, if you find the secretory type, you have a different mucosa, which I do not think is comparable with the findings of Dr Kalm.

Dr Sundberg: I think that it is an established fact that we have a respiratory mucosa in the middle ear as well as in the sinus. Of course there are certain differences, but, depending on where in the middle ear you take a tissue sample.

In principle, it is the same mucosa in the middle ear as in the sinus. With goblet cells, cylindrical epithelium cells and mucus in the normal tissue. Of course, in certain parts of the middle ear, the cells are not so cylindrical and have a more cubic form, but very near to the opening of the eustachian tube, you have these epithelia. During a secretive otitis media, there is round-cell infiltration, an increase in the number of goblet cells, and an increase of the mucus secretion. You see the same in the chronic sinusitis. So, I mean that there is a same principle in the action of this respiratory mucosa.

Chairman: I think we must be coming to an end. I just wonder if Dr Mattie would like to make some concluding remarks regarding what we have learned this afternoon.

Dr Mattie: When I gave my paper at the beginning of this afternoon, I only could speculate about the papers that were going to be given this afternoon. I think, on the whole, they gave a good picture of the known data. The problems in pharmacokinetics are always the interpretation of the data, and the semantics. One of the problems in the literature–and I think this came up this afternoon– is that authors are trying to do two things at the same time, thereby confusing pharmacokinetics with therapeutic efficacy. For example, the parallel with the urinary-tract infections is well-known. If one wants to know how rapidly the drug is cleared by the kidneys, then its renal clearance should be measured. This is done by collecting the urine, determining the concentration, multiplying that by the volume to know what the kidney really has done–that is pharmacokinetics. If, on the other hand one wants to know whether the drug is effective in urinary-tract infections, the patient should not drink too much in order to obtain high concentrations, so that it can be demonstrated that the drug is an effective drug for the treatment of urinary-tract infections; that is pharmaco-therapeutics and not pharmacokinetics.

I still think that the interpretation of the data on concentrations that we have seen this afternoon are very difficult to interpret, because the compartments are not as accessible as, for instance, the urine is. We really do not know how much sputum, how much bronchial secretions are produced and how much they are diluted. One may imagine that if the drug enters very easily into the secretory cells, the drug will be secreted together with the secretion into the sputum. If it is not–which I suspect of some drugs–this gives a different picture. Therefore it is very difficult to tell which is the better-penetrating drug. In performing these kinds of experiments, we should keep all this in mind, so that we will be able to decide whether it would be better to do an animal experiment, or collect small samples of sputum of patients.

CLINICAL USE OF ANTIMICROBIAL AGENTS

11. ANTIBIOTIC TREATMENT OF SINUSITIS AND OTITIS

P.B. VAN CAUWENBERGE

INTRODUCTION

Many cases of otitis and paranasal sinusitis do not require any antibiotic treatment because their clinical picture is mild and spontaneous recovery is noted after a few days. There are also cases of sinusitis and otitis where antibiotic treatment is superfluous, useless and insufficient.

Nevertheless we must say that the majority of patients suffering from sinusitis and otitis do receive an antibiotic treatment sometime in the course of their infection. In general they really need it. I will discuss in this report the physio-pathological mechanisms leading to sinusitis and otitis, the role of the bacteria in both infections, the principles of the treatment with special reference to the antimicrobial treatment.

PHYSIOPATHOLOGY OF SINUSITIS

It is a well documented fact that the direct cause of sinusitis and otitis is generally not a bacterial one. Only in rare cases can a direct infection of the sinusal and middle-ear mucosa be shown [1]. The physiopathology of the initial stages of otitis media and sinusitis is very similar. The paranasal sinuses and the middle ear cleft–which can be considered our fifth sinus–are air-containing cavities enclosed within rigid bony walls. The tympanic membrane is the only non-bony barrier. The common feature of these cavities is that they are in communication with the nasal cavity and that this communication is a rather narrow one, which–even in non-pathologic conditions–is hardly large enough to provide a proper ventilation of the cavities. A good ventilation however is a necessity for a normal functioning of the mucous membranes of paranasal sinuses and middle ear. Both frontal sinuses have a long-drawn ostium, the ductus naso-frontalis; the common sfenoidal sinus has generally one ostium at its front wall; the ethmoidal sinuses have one opening for the anterior cells and one for the posterior group, while the maxillary sinus often presents one or more accessory ostia besides the main ostium. The diameter of each sinusal ostium is not larger than 3–4 mm. The bony ostia are covered by a mucosa with a respiratory epithelium.

Ventilation of the middle ear is provided by the Eustachian tube, a 5 cm long tube with a partly cartilaginous and a partly bony wall and covered by a respiratory mucosa. The Eustachian tube is a virtual cleft because it only opens during swallowing by a contraction of the M. levator veli palatini. There is always consequently in resting conditions a slight negative pressure in middle ear and mastoid, which is converted into an ambient air pressure when we swallow.

All kinds of nasal pathology causing edema of the nasal mucous membranes may also provoke edema of the mucosa of the sinusal ostia. The most common cause of this edema is a viral infection of the nasal mucosa. Bacterial rhinitis occurs nearly only as a secondary infection of an initially viral rhinitis. Other nasal conditions such as allergic or vasomotor rhinopathy, nasal polyposis, endonasal tumors, bad atmospheric conditions, septal deviations and other morphologic anomalies may cause a swelling or blockage of the ostial or tubal mucosa [2].

When the ventilation of the air-containing cavities is impaired by a total or partial obstruction of the ostia, hypoxygenation of the cavity is the result. In this stage a transudation occurs at the mucosa, as a result of a vasodilation of the subepithelial capillaries. This stage is still a non-infectious or sterile stage; only inflammation is noted. Due to the obstruction of the ostium, the transudate will stagnate in the sinus or in the middle ear, and as is the rule for all stagnating fluids in the body, it will become surinfected by bacteria. From now on we are dealing with a purulent sinusitis or otitis. This development of inflammation and infection is the rule for sinusitis and for most cases of acute otitis media. Nevertheless we are all aware of the existence of dentogenous sinusitis and of chronic secretory otitis media, in which this mechanism of inflammation and infection is not valid. To go into details concerning these subjects will lead us too far from the subject of this symposium.

THE ROLE OF BACTERIA

In the majority of cases of purulent sinusitis bacteria are present. Although reports dealing with bacteriological findings in sinusitis are characterized by a diversity of results, most authors agree that purulent sinusitis secretions seldom give sterile cultures [3, 4, 5, 6]. The sterile culture rate ranges from 0% [6] to 33% [5]. My own results give a positive culture in 84% in acute sinusitis and 83% in chronic sinusitis [4]. Yet even obtaining a sterile culture does not necessarily mean that bacteria are not involved in the sinusitis process, because Brorson et al. [7] found serological evidence of bacterial infection in patients where the cultures remained sterile. We can state that with better methods for bacteriologic examination (transport medium, anaerobic cultures and serological investigations), nearly all sinusal secretions harbour bacteria. Since normal sinuses–

contrary to the nasal cavity–are germ-free, we may suggest that bacteria really play a role somewhere in the course of sinusitis.

When we consider the different bacteria involved we must emphasize that the bacterial spectrum of sinusitis today is not the same as in earlier days [4]. For example, at present we rarely find Streptococcus pyogenes in the sinusal secretions, while in the pre-antibiotic era these bacteria were very common in sinusitis [3]. Considering the so-called pathogenic aerobes we know that nowadays Streptococcus pneumoniae is the most common bacterium; we found it in 34% of all cases of acute sinusitis. It is followed by Haemophilus influenzae (19%), Staphylococcus aureus (13%), and Streptococcus pyogenes (7%). In chronic sinusitis Haemophilus influenzae and Streptococcus pneumoniae are present in respectively 18% and 17%, Staphylococcus aureus in 12%. In acute sinusitis gram-negative rods are very seldom cultured, in the chronic forms Pseudomonas, Proteus and Klebsiella are sometimes encountered. (Table 1) [4].

Until now I have only mentioned the so-called pathogenic bacteria, but I am convinced that the assumed non-pathogenic bacteria are not so innocent; they belong to the normal flora of the nasal cavity, but they do not belong to the

Table 1. Occurrence of presumed pathogenic aerobic bacteria in acute and chronic sinusitis*.

	Acute sinusitis (n = 69)		Chronic sinusitis (n = 181)	
	(n)	(%)	(n)	(%)
Streptococcus pneumoniae	24	34	30	17
Haemophilus influenzae	13	19	33	18
Staphylococcus aureaus	9	13	21	12
β-haemolytic streptococcus	5	7	4	2
Pseudomonas species	3	4	12	7
Klebsiella species			10	6
Proteus species			6	3
Escherichia coli			2	1

* From van Cauwenberge et al., 1976 [4].

Table 2. Occurrence of presumed non-pathogenic aerobic bacteria in cases of sinusitis*.

	(n)	(%)
Staphylococcus epidermidis	49	14
α-haemolytic streptococcus	32	10
γ-haemolytic streptococcus	25	7
Corynebacterium pseudodiphteriteum	13	4
Neisseria catarrhalis	10	3

* From van Cauwenberge et al., 1976 [4].

normal sinusal flora, because there is no bacterial flora in the normal sinus. Staphylococcus epidermidis is found in 14% of all secretions, the alpha-hemolytic streptococci in 10%, the gamma-hemolytic streptococci in 7% and Branhamella catarrhalis in 3%. (Table 2) [4]. The role of Branhamella however is certainly more important than its presence in the secretions suggests, because Brorson et al. demonstrated in 25 patients out of 97 a significant titre change of complement-fixing antibodies to these bacteria, while in only 3 cases Branhamella was found in the cultures of the secretions [7].

Due to the difficulty of culturing anaerobes, and because of the very strict measures that have to be taken to keep the material fit for anaerobic examination, there were until 1974 no reliable studies published about the role of anaerobes in sinusitis. In 1974 Frederick and Braude [8] demonstrated anaerobic growth in 43 cases out of 83 (52%). Their specimens were aseptically removed from chronically infected paranasal sinuses during radical sinus surgery. In 23 cases they found a pure heavy growth of anaerobes. (Table 3) [5, 6].

Our studies [4, 9, 10] include out-patients suffering from acute and chronic sinusitis. We found a pure aerobic culture in 50%, a mixed aerobic-anaerobic flora in 21%, a pure anaerobic culture in 12% and sterile cultures in 17%. Thus, anaerobes were present in 33% of the cases. In our studies Peptostreptococcus had the highest occurrence of 15%, followed by Peptococcus, Veillonella, Propionibacterium acnes, all found in 7%, and Bacteroides spp. 5%. Frederick and Braude found more Bacteroides than I did, but their patients all suffered from a long lasting chronic sinusitis.

It is not surprising to find this high number of anaerobes in sinusitis, considering the favourable conditions in the infected sinusal cavity. Indeed, the decreased

Table 3. Anaerobic infection of the sinuses.

83 Patients with chronic sinusitis*			100 Sinus secretions from 66 out-patients with acute or chronic sinusitis**	
	Number of samples°			Number of samples°°
Aerobes only	19	(23%)	Aerobes only	50
Anaerobes only	26	43 (52%)	Anaerobes only	12
Mixed aerobes-anaerobes	17		Mixed aerobes-anaerobes	21 33
No growth	21	(25%)	No growth	17
Heavy growth of anaerobes in 23 cases		(28%)		

* From Frederick and Braude, 1974 [8].
** From van Causenberge et al., 1975 [9].
° Samples obtained during surgical procedures.
°° Samples obtained by antral puncture.

mucosal blood flow caused by the insufficient drainage, the increased intrasinal pressure and the occasional angiitis, together with the presence of viscid secretions, provoke a low oxygen tension and a low pH and consequently the optimum oxydation-reduction potential necessary for anaerobic proliferation [10]. In particular unilateral sinusitis, dentogenous sinusitis, very severe infections and those presenting putrid secretions are likely to contain anaerobes [10].

The role of bacteria in acute otitis is very similar to that in acute sinusitis. It is best demonstrated by the study of Brook [11] who found Streptococcus pneumoniae in 37%, Haemophilus influenzae in 30%, Staphylococcus aureus in 9%, Streptococcus pyogenes in 5%, plus Staphylococcus epidermidis in 5%. These figures concerning acute otitis media in the U.S.A. and published in 1979 are almost the same as our figures concerning acute sinusitis in Belgium, published in 1976.

Brook also found a pure aerobic culture in 13% and a mixed aerobic and anaerobic flora in 14% of the acute otitis media cases. Contrary to our findings in sinusitis, Peptococcus was more often found than Peptostreptococcus (17% and 4% respectively).

In chronic otitis media we find a completely different bacterial spectrum. We shall not discuss this subject because there is no need for antibiotic treatment in this kind of infection. Pseudomonas aeruginosa, Proteus species and Staphylococcus aureus are by far the most common bacteria occuring in chronic otitis media. Anaerobes are also found in one third of the cases.

PRINCIPLES OF TREATMENT IN SINUSITIS AND OTITIS

Sinusitis

The histopathologic changes in acute sinusitis are reversible lesions, such as vasodilation of the capillaries of the lamina propria, edema and infiltration of the submucosa and necrosis of the epithelium [1]. The treatment of an acute sinusitis is consequently a conservative one. It is based upon two principles: (1) we must restore a normal ventilation of the sinus, and (2) in cases of purulent sinusitis a proper antibacterial treatment should be given.

The sinusal ventilation may be restored by the administration of local or systemic vasoconstricting drugs or anti-inflammatory compounds. Antibiotics are indicated when the purulent stage is reached. It seems unnecessary to start earlier with this treatment because several kinds of sinusitis heal spontaneously before the purulent stage is reached and because we found that the 'profylactic' administration does not have any influence on the possible development of a purulent sinusitis.

In subacute sinusitis–lasting 3 weeks to 3 months–the histological picture shows the proliferation of young, well vascularized connective tissue. The physio-pathological process is a battle between the aggression of micro-organisms and the defense of the sinusal mucosa [1]. The treatment should be focused on helping our defense mechanisms by eliminating the aggressors. The antral irrigation must be performed; it removes the infected contents of the sinus and opens the sinusal ostium. If we find purulent secretions, an antimicrobial treatment should be instituted. Most E.N.T. practitioners, including myself, administer antibiotics in the sinusal cavity after the irrigation. It is not proven that this is of any use, however we feel safer. We have the impression that if a viscid solution is used for local application in the sinus, the development of a sinusal mycosis is promoted.

The histopathologic changes of chronic sinusitis are characterized by chronic–mostly irreversible–lesions: necrosis and proliferation. All layers of the sinusal mucous membranes may take part in this process, even the periosteum and the bone [1]. After an attempt to cure the disease with sinusal irrigations and correction of the predisposing factors, we often have to resort to surgical methods. This surgical help consists of two important acts: (1) the removal of all irreversibly damaged tissue, and (2) the restoration of a proper sinusal ventilation.

The indication for sinusal surgery depends on the degree of symptoms the patient presents; when the only symptom of the chronic sinusitis is a slight mucous rhinorrhea, we are not in a hurry to expose the patient to a general anesthesia and possible postoperative complications such as hypesthesia or neuralgia of the infraorbital region. We must weigh the expected advantages against the possible inconveniences and side effects.

The administration of systemic antibiotics is useless because the main process is not an infection, but irreversible tissue lesions, mostly combined with an impaired ventilation or predisposing factors.

Otitis

The same principles of sinusitis treatment are also valid in otitis. In acute otitis media we must restore a good drainage and ventilation. This can be archieved by performing an incision of the eardrum–a myringotomy–which will allow the middle-ear secretions to leave the cavity and allow the ambient air to enter the unventilated middle-ear cleft. We can also administer local or systemic vaso-constricting agents.

When the acute otitis media reaches the stage of transudation or exudation, and if no tympanocenthesis is performed, we should start an antibiotic treatment to cure the infection before a spontaneous perforation of the drum occurs. If the tympanocenthesis reveals a purulent content of the middle ear, antibiotics are also indicated to prevent a long-lasting history of purulent aural discharge.

The treatment of a spontaneously perforated drum caused by an acute otitis media consists therefore of the aspiration of the pus and of the administration of the proper antibiotic.

The treatment of chronic secretory otitis media and of chronic suppurative otitis media does not include systemic antibiotic treatment. In chronic secretory otitis media, restoration of the impaired middle-ear ventilation is the most important therapeutical measure; the insertion of transtympanic ventilation tubes here is the most frequently applied technique.

In chronic suppurative otitis media, we very often note the presence of a cholesteatoma or inflammatory polyps. Careful aspiration of these lesions under microscopic view or surgical removal are indicated here. Systemic antibiotic treatment is of no use, while the topical administration of antibiotic only makes sense in treating acute episodes of infection occurring in this chronically diseased tissue.

THE ANTIMICROBIAL TREATMENT OF SINUSITIS AND OTITIS

It has already been mentioned that systemic antibiotic treatment is very useful in the treatment of the acute and subacute stages of sinusitis and otitis. In the chronic stages they are inadequate and thus superfluous.

Antibiotic treatment in cases of sinusitis must continue for at least 10 days. The antimicrobial drug of choice must satisfy several conditions: (1) it must have sufficient antibacterial activity against the bacteria most frequently involved in acute sinusitis and otitis: (2) it must have adequate sinus or middle-ear tissue penetration; and (3) its side effects must be minimal.

For the treatment of an uncomplicated sinusitis and otitis we can easily exclude two groups of antibiotics because of their important possible side effects: the aminoglycosides and chloramphenicol. The usual benign evolution of a sinusitis or an otitis does not allow the administration of antibiotics with nephro- and oto-toxic side effects or causing an irreversible depression of the hematopoietic system.

Due to their insufficient antibacterial spectrum we can also exclude penicillin G and the semi-synthetic penicillinase-resistant beta-lactam drugs (oxa-, cloxa-, dicloxa- and flucloxa-cillin). We readily accept that a first choice antibiotic in the treatment of sinusitis and otitis must be active against all bacteria that can be found in the majority of cases. These bacteria are Streptococcus pneumoniae, Haemophilus influenzae, Branhamella catarrhalis, Peptostreptoccus (or Peptococcus for otitis), Staphylococcus aureus, Staphylococcus epidermidis and the alpha-haemolytic streptococci. Penicillin is not active enough against H. influenzae and the vast majority of Staphylococcus aureus; the penicillinase resistant beta-lactam drugs are inactive against H. influenzae, while their resorption after oral administration is variable, and the M.I.C. for most bacteria is higher than for the other betalactamines.

The remaining antibiotic groups all have a sufficient penetration in the sinusal mucosa [12]. This is demonstrated for amoxycillin, bac- and piv-ampicillin (better than ampicillin), several cephalosporins, the macrolides (erythromycin and spiramycin), clindamycin and the tetracyclins (doxycyclin, lymecyclin and minocyclin). The sinusal tissue concentration of penicillin is poor, and is still not known for cotrimoxazol. In view of these facts we can restrain some antibiotics or groups of antibiotics that have a place in the treatment of sinusitis and otitis; they all have some inconveniences, none is perfect but they are useful in the great majority of cases. Taking into account their respective advantages and inconveniences, we can select one drug for the treatment of our individual cases.

The ampicillin group, including the esters and amoxycillin, has a good spectrum. It is, however, not active against most strains of Staphylococcus aureus. The spectrum is broad and includes several gram-negative bacteria which are not important in sinusitis and otitis. The esters of ampicillin and amoxycillin have a better sinusal tissue concentration than ampicillin itself. Concerning the inconveniences, we can state that ampicillin causes fairly often some diarrhea because of the poor absorption in the gastro-intestinal tract and its broad spectrum; this side effect is nearly absent for amoxy-cillin and ampicillin-esters. Pivampicillin often causes stomach ache because formaldehyde is one of the metabolites; because of this side-effect this drug is not promoted anymore in some countries. The recommended doses for adults are: ampicillin 500 mg q.i.d., bacampicillin and talampicillin 400 mg t.i.d. and amoxycillin 350 mg q.i.d. or 500 mg t.i.d. or q.i.d.

Although the cephalosporines have a good (but broad) spectrum and the sinusal tissue concentration is good, I do not consider them as a first choice antibiotic in the treatment of a common purulent sinusitis and otitis because they are still too expensive. If there are indications, for example, because of the sensitivity of the isolated micro-organism, the recommended dosage is 500 mg q.i.d. The macrolides (erythromycin, spiramycin, troleandomycin) have a good antibacterial spectrum except for the rather poor activity against H. influenzae. The spectrum is not too broad and the macrolides seldom disturb the normal flora. The sinusal tissue concentration of erythromycin and spiramycin is good. With the recommended dose of 500 mg q.i.d. we may sometimes note slight stomach ache.

Lincomycin and clindamycin are usually classified in the group of macrolides; the pharmacological properties do not differ much from erythromycin and company except for the good activity against Bacteroides fragilis, which, however, are seldom found in sinusitis and otitis. Clindamycin may cause a muco-membranous colitis.

The tetracyclins, doxycyclin, minocyclin and lymecyclin all have a good spectrum, but we are all aware of the increasing resistance of Pneumococci and H. influenzae. Their sinusal tissue concentration is good. They may cause a discoloration of the teeth when given to young children (doxycyclin and minocyclin in a much lesser degree than the older tetracyclins) and minocyclin often causes dizziness because of excellent passage through the blood-brain barrier. The recommended dosages for doxycyclin and minocyclin are 100 mg b.i.d. or once a day after an initial gift of 200 mg. Lymecyclin is administered 300 mg b.i.d.

For the treatment of severe or complicated cases of sinusitis and otitis, surgical drainage of the sinusal, middle-ear or mastoid empyema is required. Here, the parenteral administration of high doses of antibiotics is recommended. Cephalosporines, doxycyclin, a combination of ampicillin and dicloxacillin and even chloramphenicol are indicated.

CONCLUSION

Most cases of sinusitis and otitis not only require a restoration of the proper ventilation and drainage of the infected cavities, but also an antimicrobial treatment. Antibiotics are indicated because bacteria are present in the majority of the purulent sinusal and middle-ear secretions, because the healing process is faster when antibiotics are administered, and because complications can be prevented.

The antimicrobial drug of choice should have (1) a good activity against the germs encountered in sinusitis and otitis, (2) a good penetration in the sinusal tissue, and (3) minimal side effects.

REFERENCES

1. Van Cauwenberge P: Sinusitis, The Upjohn Medical Monographs, nr. 3, Puurs, Belgium, 1979.
2. Van Cauwenberge P: Anatomy and physiology of the upper airways. In: Advances in Allergology and Applied Immunology (Oehling A, Glazer I, Mathov E, Arbesman C, eds.), Pergamon Press, Oxford, 1980, p. 93.
3. Palva T, Gronroos J., Palva A: Bacteriology and pathology of chronic maxillary sinusitis. Acta Otolaryng (Stockh) 54:159, 1962.
4. Van Cauwenberge P, Verschraegen G, Van Renterghem L: Bacteriological findings in sinusitis Scand J Infect Dis Suppl 9:72, 1976.
5. Brorson J-E, Alxelsson A, Holm SE: Serological studies in acute maxillary sinusitis. Acta Otolaryng (Stockh) 82:415, 1976.
6. Carenfelt C, Lundberg C, Nord C-E, Wretlind B: Bacteriology of maxillary sinusitis in relation to quality of the retained secretion. Acta Otolaryngol 86:298, 1978.
7. Brorson JE, Axelsson A, Holm SE: Studies on Branhamella catarrhalis (Neisseria catarrhalis) with special reference to maxillary sinusitis. Scand J Infect Dis 8:151, 1976.
8. Frederick J, Braude AI: Anaerobic infection of the paransal sinuses. New Engl. J Med 290:135, 1974.
9. Van Cauwenberge P, Kluyskens P, Van Renterghem L: The importance of the anaerobic bacteria in paranasal sinusitis. Rhinology 8:141, 1975.

10. Van Cauwenberge P, Van Renterghem L, Verschraegen G, Kluyskens P: Bacteriology in sinusitis with special reference to the role of the anaerobes. International Symposium of Infection and Allergy of the Nose and Paranasal Sinuses. Scimed Publ., Tokyo, 1977, p. 151.

11. Brook I: Otitis media in children: a prospective study of aerobic and anaerobic bacteriology. Laryngoscope 89:992, 1979.

12. Axelsson A, Brorson J-E, Jensen C: Akut Sinuit. Astra Läkemedel A.B., Södertälje, 1978.

DISCUSSION

Chairman: Have you any figures how frequently sinusitis is caused by a viral infection only?

Dr Van Cauwenberge: I am afraid there are no figures available. The definition of sinusitis is very important. All kinds of nasal infection, all kinds of inflammation of the nasal mucosa may also cause inflammation of the sinusal mucosa, without reaching the stage of a real sinusitis. We can observe the same in the mastoid. When we have an acute otitis media, there is always a slight edema, an inflammation of the mastoid mucosa, without a real mastoiditis. When I speak about sinusitis, I mean the secretory type of sinusitis, where mucoid, serous or purulent secretions are present. I do not think that the real viral sinusitis does exist; we have a viral rhinitis causing a secondary sinusitis. Sinusitis caused by viruses must be extremely rare.

Dr Kalm: I should like to make a short comment on the clinical diagnosis maxillary sinusitis in relation to antibiotic treatment. There have been very few studies on relationships between the so-called clinical picture and the radiological findings. The symptomatology described in conventional textbooks in very unspecific and often there is no reference to controlled investigations. In Sweden, Axelsson and co-workers published a study done in 164 patients suspected by either the patient himself or one of three otolaryngologists of having sinusitis.

Table 1. Symptoms positively correlated to sinusitis*.

Radiological point values	0 (%)	1–3 (%)	4–6 (%)
URI precedings complaints	67	85	90
Nasal discharge	65	82	97
Nasal discharge, thick or purulent	44	65	75
General malaise	42	55	60
Gough (bronchitis)	42	64	75
Hyposmia	40	61	75
Fever exceeding 38 °C	2	14	28

* From ORL, 1976, 38, 298.

These patients were asked a series of questions and examined clinically. An X-ray examination was performed, and the pathological findings were graded on a 0–6 point scale according to their severity.

It can be seen from Table 1, that the symptoms correlated positively with the radiological findings were very unspecific and in many cases almost impossible to distinguish from an uncomplicated rhinitis. The best correlation with symptoms was found for (purulent) nasal discharge, general malaise, coughing, hyposmia, and pains at mastication.

There was a negative correlation which is perhaps a bit surprising between the X-ray findings and such symptoms as frontal or retro-ocular pains, ear pressure, and foul nasal smell (Table 2). Among the symptoms showing no correlation, positive or negative, with the radiological findings were nasal obstruction, sinus tenderness, pain on leaning forward or on walking, and pressure tenderness of the upper teeth or over maxillary or frontal sinuses.

Despite these findings the diagnostic performance of the otolaryngologists was fairly good. About 70% of the cases without radiological changes were correctly predicted, but this percentage was much lower for sinuses with radiological changes. Total opacity of the maxillary sinuses was predicted in only 43% of the cases. On the basis of these results we think it advisable to start with a narrow-spectrum antibiotic if the diagnosis sinusitis is based solely on the clinical picture. The chance of incorrect diagnosis and overtreatment of sinusitis seems to be great if sinus puncture or radiological examination is not performed routinely.

Dr Van Cauwenberge: I agree.

Dr Simon: Do you see any difference in the speed in improvement between bactericidal and bacteriostatic antibiotics?

Dr Van Cauwenberge: I sometimes have the impression that bactericidal antibiotics are more active. Not that the disease will be cured earlier, but I have the impression that more cases are cured. However, there have been studies published comparing doxycycline and amoxicillin and doxycycline versus bacampicillin and there were no significant differences.

Table 2. Symptoms and signs negatively correlated to sinusitis*.

Radiological point values	0 (%)	1–3 (%)	4–6 (%)
Frontal pains	72	46	42
Previous sinusitis	67	59	47
Ear pressure	38	25	20
Subjective foul nasal smell	33	30	18
Retro-ocular pains	28	14	8

* From ORL 38: 298, 1976.

Dr Gould: I was very interested to see Dr Van Cauwenberge's figures on the anaerobes and I would suspect that aerobes are certainly present in combination with facultative anaerobes in many more cases, if care is taken to look for them. However, you did mention that 12% of your cases had a pure anaerobic culture. I wanted to ask if you had any experience in using drugs specific against anaerobes, such as metronidazol for example, in the treatment of sinusitis.

Dr Van Cauwenberge: No. I think sinusitis and otitis media are not such bad infections as to require specific anti-anaerobic treatment with metronidazol. It is better to do a sinal puncture and irrigation, to restore the intrasinal milieu.

Chairman: To follow this up, did I understand clearly: when you do sinus irrigation, do you also leave antibiotics locally, or is it only in special cases?

Dr Van Cauwenberge: No, I do it routinely. After the irrigation, I inject an antibiotic in the sinus.

Chairman: What antibiotic did you select for this?

Dr Van Cauwenberge: There are several possibilities. I used thiamphenicol and erythromycin (the intravenous form). Now I use chiefly thiamphenicol, because there is also a mucolitic agent in that preparation. I do not know if it is active, I just feel more secure when I do it.

Chairman: Do you want to advocate it or not?

Dr Van Cauwenberge: No, I do not want to advocate it.

Dr Van der Meer: If you give thiamphenicol locally and you give a systemic therapy, for instance with ampicillin, you might come into some drug interaction. What is your idea about that?

Dr Van Cauwenberge: I do not give any systemic antibiotic treatment in chronic sinusitis. Most of the cases in which I performed sinus irrigation are chronic forms of sinusitis in which I do not give any systemic antibiotic treatment.

Dr Butzler: You mentioned two grams of erythromycin. I suppose it is the stearate compound. We will never prescribe two grams of stearate, because it gives you such a high percentage of intolerance.

Dr Van Cauwenberge: In the last few years, I have used three times 500 mg more often than four times 500 mg, but if patients are not complaining, I continue giving four times 500 mg.

Dr Butzler: It is our experience that if you give four times 500 mg, you will have about 10 to 15% treatments to stop.

Dr Van der Waaij: May I come back to the end of your lecture regarding what you leave behind in the sinus. It is not better to use a non-absorbable antibiotic? Secondly: do you investigate the purulent discharge for the presence of granulocytes as a sign of inflammation?

Dr Van Cauwenberge: Yes, we always make a direct microscopic examination to look for granulocytes. This gives us a very good idea of what stage we are in, whether it is really an active inflammatory stage, or whether there are chronic lesions without real infection, just polyps that block the sinus. The resorption of antibiotics from the infected sinusal mucosa is very poor, according to some studies done in Japan. The nasal mucous membranes are known to be very good absorbers of all kinds of drugs, but this is not the fact for the infected sinusal mucosa.

Dr Van der Waaij: If you use a small spectrum, you may have just a change in the flora that lives happily in the sinus. I would use a broad spectrum antibiotic locally and a non-absorbable one, if it exists.

Dr Stam: I would like to come back to that point, too. I think it is useless to leave behind an antibiotic in the sinus. Where is the infection? The infection is not in the discharge; the infection is in the mucosal wall. So, if you use something that does not reach the wall, it is useless. We did it with pulmonary abscesses and now we know that some antibiotics penetrate the pleural fluid, the lung tissue, the ear and the sinuses very well. So, you might just as well forget it.

Dr Van Cauwenberge: C. Lundberg et al. recently provided evidence that bacterial multiplication starts in the sinal secretion and that the bacteria enter at an early stage of the infection into the goblet cells. They also proved that before any true bacterial invasion of the mucosa occurs, a clinically purulent infection is established. Bacterial invasion of the subepithelial layers will not take place until the normal arrangement of the epithelial cells is disrupted. These findings may be in favor of a local application of antibiotics (or antiseptics) in the sinal cavities.

Dr Mouton: I would like to come back to the matter of the anaerobic infections. I think the question which was raised by Dr Gould has some other significance. I am very often in doubt about the significance of anaerobes we find in sinus secretions. Undoubtedly, there are anaerobic infections in the sinus, but I wonder, when you find anaerobes, whether you are dealing with a mixed infection.

Dr Van Cauwenberge: I think there are three different kinds of anaerobic infections. First, the real anaerobic infections where the anaerobic bacteria are at the origin of the infection. These are rare. The largest group is that of chronic sinusitis, where the anaerobes are just contaminants who found an ideal growth medium in the chronically infected sinus cavity. The third group is that of the dentogenous sinusitis, where in apical abscesses and granulomas, anaerobes are by far the most common bacteria. When you have a spreading of this dental infection to the sinusal cavity, this is very hard to cure.

Dr Sundberg: I would just ask you a question concerning the etiological agents in acute otitis media. You mentioned that Staphylococcus aureus is one of the pathogens, and I know many share that opinion. If you use a careful aspiration technique when you are sampling middle-ear exudate, it is almost impossible to obtain strains of Staphylococcus aureus. My question is: is Staphylococcus aureus just a contamination, or is it an etiological agent?

Dr Van Cauwenberge: I think it is very seldom an etiological agent. I only found about 3 to 4% Staphylococcus aureus, and even in these cases, I am not sure they really were the etiological agent. As you know, Staph.epidermidis and Staph.aureus are very common in the outer-ear canal as well as in the nasal mucosa.

Dr Van Marion: We reported some years ago on three cases of reversible hearing loss during erythromycin therapy. We should like to ask if anyone from the audience has a suggestion concerning the mechanism whereby the loss of hearing could have been caused.

Dr Grote: There have been several cases reported about hearing loss due to erythromycin. Have you done further audiological investigations regarding the type of hearing loss it was, a cochlear or retro-cochlear hearing-loss type?

Dr Van Marion: One of our patients in the ear-nose-throat department. They thought there was a cochlear type of hearing loss.

Chairman: Is there anybody else who has had this type of experience? Apparently not.

Dr Thompson: In addition, I can say that we had to discontinue erythromycin therapy in two patients with endocarditis, also because of impaired hearing.

Chairman: Thank you. Apparently, this is a Leiden disease at the moment.

12. ANTIBIOTIC TREATMENT OF CHRONIC BRONCHITIS

R.J. DAVIES and G.K. KNOWLES

CHRONIC BRONCHITIS

Chronic bronchitis remains a major cause of disability in the middle aged and elderly and is responsible for approximately 20000 deaths each year in the United Kingdom. Despite the importance of cigarette smoking in the aetiology of this disease, environmental and social factors are also of consequence. Indeed recent information suggests that both disability and death from chronic bronchitis are decreasing in the U.K. possibly as a result of the clean air policy which has dramatically reduced the level of atmospheric pollution [1]. The prevalence of chronic bronchitis varies greatly in different countries independent of cigarette smoking, and between rural and urban areas [2]. However, in one recent study of over 12 000 individuals aged between 45 and 70 years old in South East London, symptoms of chronic bronchitis were reported in 15% [3].

The fundamental disorder in chronic bronchitis is hypertrophy and hyper-activity of the mucus secreting glands and goblet cells throughout the bronchial tree and when mucus hypersecretion is the only clinical abnormality, the British Medical Research Council's Committee on Bronchitis classified this disease as 'simple bronchitis'. They also described two other types of chronic bronchitis one characterized by the persistent or intermittent presence or pus in the sputum–chronic or recurrent mucopurulent bronchitis, and the other associated with persistent widespread narrowing of the bronchial airways at least on expiration, causing increased resistance to airflow–chronic obstructive bronchitis [4]. Clinically, the most important aspect of chronic bronchitis is this structural damage particularly of small airways, which leads to progressive deterioration of lung function, severe disability from shortness of breath and death. Identification of the factors responsible for this continuing bronchial damage remains the most important problem in the understanding of this extremely common and debilitating disease.

Considerable attention has been focused on the role of respiratory tract infections in chronic bronchitis and it is universal practice to treat exacerbations of the disease with antibiotics. However, the evidence that this has any effect on the development of the irreversible airflow obstruction, so important in the symptomatology or indeed on the outcome of the exacerbations, requires clarification.

Infection in Chronic Bronchitis

Studies have suggested that only about 45% of acute exacerbations of chronic bronchitis are due to identifiable bacterial or viral infection [5, 6] and the important role of respiratory viruses in the initiation of episodes of bronchitis has been increasingly recognized. Indeed in an investigation from North America, one third of the 116 exacerbations that were studied were related to viral infections particularly with influenza and para-influenza viruses [7]. Gregg studying patients in a general practice in England, found that rhinoviruses were the commonest organism isolated during acute bronchitic episodes and showed that whereas these viruses led to the features of a 'common cold' in previously healthy individuals, patients with chronic bronchitis developed symptoms and signs of a lower respiratory tract infection [8]. Viral infection can damage the mucosa of the respiratory tract and allow secondary bacterial infection. Nevertheless, the symptomatology and treatment of acute exacerbations of this disease is dominated by bacterial infection in the lower respiratory tract, irrespective of whether this is a primary or secondary event.

Since the early 1950's many investigations have shown the association between bacteria, especially Haemophilus influenzae and Streptococcus pneumoniae, sputum purulence and exacerbations of chronic bronchitis. May noted this association in sputum and suggested that appropriate antibiotic treatment should eradicate the bacteria and clear the purulence [9]. Samples taken directly from the trachea of non-bronchitic controls and patients in hospital with bronchitis showed that whereas the lower respiratory tract of the controls was generally sterile, H. influenzae and Strep. pneumoniae were frequently isolated from the patients with bronchitis [10, 11]. Studies of precipitating antibodies in serum against these organisms have in general confirmed these findings [12]. However, difficulties remain in establishing a clear cut role for H. influenzae and Strep. pneumoniae in the aetiology of exacerbations of chronic bronchitis and in part this is due to the widespread colonisation of the upper respiratory tract of 'normal' asymptomatic individuals by these organisms. Indeed H. influenzae and Strep. pneumoniae can be cultured from sputum in almost 50% of chronic bronchitics expectorating mucoid sputum [13]. Sequential studies using such techniques as transtracheal aspiration to avoid contamination of sputa by oropharyngeal bacteria in patients with chronic bronchitis during both exacerbations and remissions are required to elucidate this problem.

Antibiotic Treatment

Antibiotics have been widely used for the treatment of acute exacerbations of chronic bronchitis associated with expectoration of purulent sputum. However, some doubt remains as to whether such treatment significantly influences the outcome. Tagar and Speizer [14] reviewed six studies in which the effect of antibiotic therapy was compared with placebo in the management of acute exacerbations of bronchitis. The antibiotics included oxytetracycline, ampicillin, chloramphenicol and penicillin and streptomycin in combination. They concluded that four of the six studies showed no significant benefit in favour of the group receiving antibiotics. The results obtained from the other two studies, which claimed a beneficial effect for antibiotic treatment, were open to criticism in one case on grounds of statistical interpretation and in the other the benefit of antibiotic treatment rapidly disappeared and most of the patients became ill again with 7–28 days [15].

The effect of longterm administration of antibiotics to patients with chronic bronchitis has in general proved of little value in preventing exacerbations. Large scale controlled studies by Francis and co-workers [16] showed that tetracycline in a twice daily dose of 250 mg decreased the number of days lost from work from acute episodes of bronchitis, but did not decrease the frequency of exacerbations. Controlled trials by the Medical Research Council of Great Britain [17] in which higher doses of tetracycline were administered showed that benefit was limited to a decrease in the number of exacerbations in bronchitics who had frequent relapses. Again the number of days lost from work was lower in the tetracycline treated group but even this effect was open to question on statistical grounds.

Despite this evidence, antibiotics are widely prescribed for acute exacerbations of bronchitis with expectoration of purulent sputum, and few clinicians would withhold this form of therapy. A large number of trials have been performed in which different antibiotics have been compared in the treatment of acute episodes of bronchitis. When allowance is made for variations in the criteria for case selection and differences in dosage schedules, no single drug has emerged as clearly superior to the rest in term of therapeutic efficacy or freedom from undesirable side effects. Ampicillin in a dose of 4 g per day may be slightly more effective than the tetracyclines [18] and in a single blind comparative study, co-trimoxazole was shown to be more effective than ampicillin at least in terms of reduction in sputum volume and purulence. In this study, as in others, antibiotic treatment did not always lead to clearance of purulent sputum and treatment success was considered as a change from purulent to predominantly mucoid sputum on visual inspection. The only other variable considered was the clinician's overall assessment [19]. Cephalosporines are probably of no greater benefit than ampicillin [20]. Amoxycillin is closely related chemically to ampicillin and in vitro both substances display substantially the same antibacterial activity. In vivo, amoxycillin is absorbed twice as efficiently as ampicillin giving double the serum concentration and studies by May and Ingold [21] suggested that 500 mg of amoxycillin 6 hourly was more effective in clearing sputum purulence that 1 g of ampicillin 6 hourly. Further they produced some evidence to suggest that amoxycillin might penetrate the bronchial mucus membrane more easily than ampicillin. Some strains of H. influenzae show relative resistance in vitro to erythromycin, and this drug has rarely been used in the treatment of chronic bronchitis. However, in an out-patient of study of 72 patients with exacerbations of chronic bronchitis, erythromycin and ampicillin each at a dose of 500 mg four times a day were found to be equally effective [22].

Comparison of Amoxycillin, Erythromycin and Co-trimoxazole in Exacerbations of Chronic Bronchitis
These three antibiotics were compared in a random order study which of necessity could only be single blind due to the different formulation and dose schedules of the drugs.

Patients. Fifty five patients with acute exacerbations of chronic bronchitis severe enough to require hospital admission entered into the comparative trial. Six of the patients were admitted twice during the period of the investigation giving a total of 61 acute exacerbations of chronic bronchitis for study. The criteria for admission into the trial were that the patients had to have chronic muco-purulent bronchitis with at least two exacerbations requiring treatment in the previous year and no underlying malignancy or pneumonia. No patient with a history of allergic reactions to any of the antibiotics was included and all showed less than 15% improvement in peak expiratory flow rate (PEFR) forced expiratory volume in 1 second (FEV_1) and forced vital capacity (FVC) after administration of two puffs of salbutamol (200 μg). The study patients had evidence of very severe chronic obstructive bronchitis since the mean FEV_1 for the group was only 30% of their predicted values: 43 were males and 12 females and the mean age was 64.4 ± 1.4 yr (mean $\pm$ standard error). All patients were skin prick tested with extracts from 23 common environmental and food allergens and 25% were found to be atopic on the basis of a positive reaction to 1 or more of the extracts. This proportion of atopic individuals is exactly that found in a control population of similar age[23]. Twenty six of the patients were current cigarette smokers, 23 had given up smoking tobacco products and interestingly 6 patients had never smoked tobacco in their lives.

Antibiotics. Amoxycillin 500 mg three times daily, erythromycin stearate 500 mg three times daily and co-trimoxazole 2 tablets twice daily were administered in random order to the patients. Each antibiotic was administered after meals and therapy was continued for 10 days.

Assessment. Daily records were kept of general state, severity of cough and breathlessness and extent of wheeze and crackles in the chest in every patient. Each of these variables was scored in a scale from 1 to 3 indicating increasing degree of severity. The sum of the daily scores for each variable provided an index of illness severity for every subject. Patients were also asked to record daily on a 10 cm visual analogue scale any improvement or deterioration in general state, cough and breathlessness. Their progress and illness severity was monitored by measurement of the total distance in centimetres from the patient's assessment points to the 'best' end of the scale for each of these variables. PEFR, FEV$_1$ and FVC were measured in triplicate on admission and discharge using a Wright's peak flow meter and Vitalograph® respectively and the highest values recorded. The results for each subject were expressed as a percentage of their predicted values [24].

Sputum was collected over two hour periods on admission and between 09.00 and 11.00 hr each day in hospital. The volume was recorded and the degree of purulence assessed macroscopically using a modification of the classification of May and May [25]. Mucopurulent sputum containing approximately 75% pus or more was graded MP + + +: when it contained approximately 50% pus it was graded MP + + and when it contained approximately 25% pus or less as MP +. Mucoid sputum (M) contained no pus. The sputum specimens were liquified and stained smears prepared using haematoxylin and chromotrope 2R [26]. The number of neutrophil polymorphonuclear leucocytes was counted microscopically and expressed as the mean number of leucocytes per 5 high powered fields. Samples of sputum obtain on admission and discharged were gram stained and appropriately cultured for bacterial growth. Untoward side effects following administration of the antibiotics were recorded.

Results and Discussion The results of this single blind comparative trial of amoxycillin, erythromycin and co-trimoxazole are shown in Table 1. The time spent in hospital was the same for the patients irrespective of which antibiotic they had received. Forty four per cent of the patients had a raised body temperature on admission and there was no significant difference in the time taken for the temperature to return to normal in the patients receiving the three different antibiotics. No antibiotic was superior in improving the illnes severity of the patient either as assessed by the physician or by the patients themselves. It is interesting to note that whereas the majority of patients felt that their hospital treatment had significantly improved their illness, the physicians felt that this had only occurred in the minority. One possible explanation for this is that the patients admitted to this trial had very severe lung disease irrespective of excerbations. Nevertheless, there was a measureable improvement in respiratory function, particularly in the forced vital capacity, over the admission period, though there were no statistical differences between the degree of improvement in the groups receiving the different antibiotics.

A highly significant overall correlation (P < 0.001) was found between sputum purulence assessed macroscopically and the mean counts per 5 high

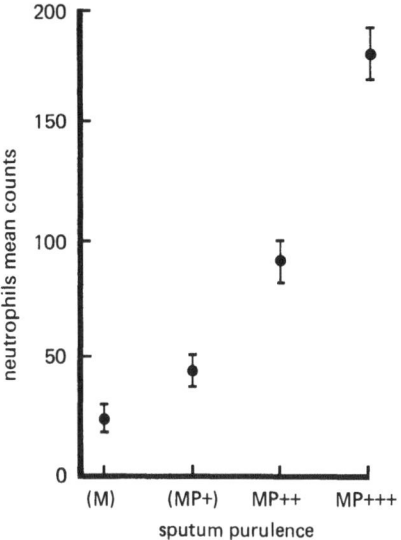

Fig. 1. The relationship between purulence, assessed macroscopically and microscopically (mean ± SE). M = mucoid sputum–MP = mucopurulent sputum; for grading see text.

powered fields of neutrophil polymorphonuclear leucocytes in liquified sputum (fig. 1). This confirms earlier reports on the value of macroscopic assessment of sputum purulence [27]. However, performing neutrophil counts in sputum does allow a more accurate assessment of the ability of antibiotics to clear sputum purulence and this is shown in Table 1. Mucoid sputum, assessed macroscopically, was only achieved by one third of the patients despite 10 days of antibiotic treatments. These results are consistent with clinical experience and indeed other studies [19] which suggest that many patients with severe chronic bronchitis continue to cough up sputum classified as containing some pus, even in 'remission'. Nevertheless neutrophil counts in sputum substantially decreased, and as with the percent of patients achieving mucoid sputum this was not significantly influenced by the particular antibiotic regimen. The majority of patients showed substantial reduction in the volume of expectorated sputum following treatment.

H. influenzae and/or Strep. pneumoniae were isolated from sputum on admission in 23 (38%) of the patients with exacerbations of chronic bronchitis. At discharge H. influenzae had been eradicated in 81% and Strep. pneumoniae in 100% of the sputa in which bacteria had previously been found. Unfortunately the number of patients from whom bacteria had been isolated in each of the three groups was too small to allow any statistical comparison of these results. H. influenzae was cultured from admission specimens significantly more frequently from patients who were current cigarette smokers (P <0.05) a result that may in

Table 1. The results of a single blind random order comparison of erythromycin stearate, cotrimoxazole and amoxycillin in the treatment of acute exacerbations in patients with severe chronic obstuctive bronchitis requiring hospital admission*.

	Erythromycin	Co-trimoxazole	Amoxycillin	Significance**
Number of exacerbations	23	23	15	
Time in hospital (days)	11.1 ± 1.5	9.4 ± 0.7	9.1 ± 1.2	NS
Time to resolution of fever (days)	1.4 ± 0.2	1.6 ± 0.3	1.5 ± 0.3	NS
Percent achieving 50% improvement in illness severity				
(A) Physician's assessment	43	39	33	NS
(B) Patient's assessment	83	70	73	NS
Percent improvement in predicted respiratory function				
PEFR	6.0 ± 1.5	8.5 ± 2.1	10.8 ± 3.6	NS
FEV_1	10.1 ± 3.2	8.8 ± 2.2	14.5 ± 3.9	NS
FVC	13.7 ± 3.8	12.3 ± 2.8	21.4 ± 4.4	NS
Percent achieving mucoid sputum	43	43	13	NS
Percent achieving 50% reduction in sputum volume	52	78	60	NS
Decrease in sputum leucocytes (number)	83 ± 33	139 ± 27	90 ± 30	NS

* Results expressed as mean ± standard error of the mean or as percent.
** The statistical significance of any differences between results in the three groups is shown.
NS = not significant.

part be explained by a recent finding that the presence of tobacco extract or pure nicotine enhances growth of H. influenzae on culture media [28].

The frequency of unwanted side effects in the three groups of patients treated with erythromycin, amoxycillin and co-trimoxazole is shown in Table 2: these occurred significantly more frequently ($P < 0.02$) in those patients treated with erythromycin. However, unwanted effects were minor and short lived in all cases and did not necessitate withdrawal or alteration of therapy.

The results of this study show that there is no difference between the effectiveness of these three antibiotics in the management of acute exacerbations of chronic bronchitis in patients with severe chronic obstructive bronchitis treated in hospital. One interpretation of these findings is that amoxycillin, co-trimoxazole and erythromycin are effective antibiotics for treatment of exacer-

Table 2. Unwanted side effects of ten days treatment with erythromycin stearate, cotrimoxazole and amoxycillin*.

Unwanted effect	Erythromycin (n = 23)	Co-trimoxazole (n = 23)	Amoxycillin (n = 15)
Indigestion	5	4	1
Nausea	7	5	1
Vomiting	4	2	0
Diarrhoea	4	0	0
Sore tongue	4	0	0
% of exacerbations with one or more side-effects	61%	30%	13%

* n = number of patients.

bations of disease in which bacteria play an important part. Alternatively it might be argued that the role played by bacterial infection is relatively insignificant or that antibiotic treatment is not necessary and hospital admission and placebo therapy might have been as effective. Carefully controlled random order double blind studies using antibiotic and placebo therapy would be required to elucidate this problem.

Relationship between Respiratory Infections, Mucus Hypersecretion and Airflow Obstruction.

There is little doubt that respiratory infections can contribute to the episodic worsening of chronic bronchitis but their relationship if any to the development of mucus hypersecretion and airflow obstruction is not as clear cut. Much of the difficulty in assessing the role of respiratory tract infection in the development of mucus hypersecretion is due to the over-riding effect of cigarette smoking. Nevertheless, Oswald et al [29] found that adult patients with chronic bronchitis had experienced more school absences because of respiratory illness and significantly more episodes of bronchitis and pneumonia in childhood compared to a matched control group. Similar results were found by Reid and Fairburn [30]. May and co-workers showed that there was an association between the presence of antibodies in serum against H. influenzae and indices of mucus hypersecretion that persisted when adjustment was made for cigarette smoking [31]. On the other hand, Fletcher and co-workers [32] showed that in men without an immediate past history of recurrent chest illnesses, the presence of sputum production led to an increased frequency of chest episodes; an observation which suggested that factors leading to mucus hypersecretion in part predisposed to the occurrence of chest illnesses rather than the other way round. Studies in children by Holland and co-workers[33] support this conclusion, since they found that children who smoke not only had more chronic cough and phlegm than their non-smoking counterparts but also experienced a greater frequency of respiratory illnesses. Although there is some evidence to suggest that frequent respiratory infections in childhood may lead to the development of mucus hypersecretion, long term prospective studies are required to test this hypothesis and at the present time the matter remains unsettled.

Longitudinal studies have shown that respiratory function deteriorates at an accelerated rate in some, but not all patients with chronic bronchitis compared with healthy non-smoking adults [33, 34]. This observed decline in respiratory function can precede the onset of sputum production. Indeed in a study of 800 workers over a period of 8 yr, Fletcher and co-workers [34] showed that a quarter of those who developed severe airflow obstruction had no sputum production at any stage, emphasizing the fact that there was no observed correlation between sputum production and the rate at which FEV_1 declined. In addition, although a relationship was noted between excessive sputum production and liability to chest infections there was apparently no causal relationship between these factors and the development of airflow obstruction. May and co-workers [31] used multiple regression analysis techniques to evaluate the correlation of known or suspected factors important in the development of chronic bronchitis with precipitins in the serum against antigens from H. influenzae. When the results were adjusted for smoking habits only sputum production and purulence, sputum eosinophilia and weight were found to correlate with the presence of H. influenzae precipitins. In particular there was no correlation between the rate of decline of ventilatory function and the presence of these precipitins in serum. The authors concluded that 'infection by H. influenzae plays no significant role in the development of airway obstruction'. Other studies support this view and have shown that once obstructive lung disease becomes manifest intercurrent illnesses have little effect on the rate of decline in FEV_1 [33]. On the basis of this information the concept has developed that chronic bronchitis may in fact represent two conditions, which though not causally related, may frequently coincide. On the one hand there is the hypersecretory syndrome which may predispose to frequent infections causing symptoms for which antibiotics are usually prescribed. On the other hand there is the obstructive syndrome which insidiously leads to deterioration in lung function and eventually death: treatment for this aspect of chronic bronchitis is unfortunately restricted to attempts to stop patients smoking cigarettes before the deterioration in lung function becomes disabling.

Although many factors have been associated with the development of chronic bronchitis, the mechanisms which lead both to the mucus hypersecretion and the airflow obstruction remain unknown. Host factors, as yet unidentified must also be of considerable importance since the majority of individuals who smoke and live in urban areas do not develop chronic obstructive bronchitis. One attractive hypothesis suggests that the abnormality lies in the alveolar macrophage which is particularly sensitive to cigarette smoke in patients who develop chronic obstructive bronchitis. Perhaps functional impairment of this cell makes it less able to neutralize bacterial toxins released by continuing low grade infection of the respiratory tract with bacteria such as H. influenzae leading to chronic inflammation in the lung airways. There is no doubt that much remains to be known about this common disease and only then will it be possible to fully assess the value of antibiotic treatment.

REFERENCES

1. Howard P: The changing face of chronic bronchitis with airways obstruction. Br Med J 2:89, 1974.
2. Reid DD: The international background to studies of chronic non-specific lung disease in children. In: Bronchitis III (Orie NGM Van der Leude R eds), Royal Van Gorcum, Assen, 1970.
3. D'Souza MF, Davies RJ, Swan AV: Factors associated with reported asthma in middle age. 1980, in press.
4. Medical Research Council: Definition and classification of chronic bronchitis for clinical and epidemiological purposes. Lancet 1:775, 1965.
5. Fisher M, Akhtar AJ, Calder MA, Moffat AJ, Stewart SM, Sealley H, Crofton JW: Pilot study of factors associated with exacerbations in chronic bronchitis. Br Med J 4:187, 1969.
6. Lambert HP, Stern M: Infective factors in exacerbations of bronchitis and asthma. Br Med J 3:323, 1972.
7. Gump DW, Phillips CA, Forsyth BR, McIntosh K, Lamborn KR and Stouch WH: Role of infection in chronic bronchitis. Am Rev Respir Dis 113:465, 1976.
8. Gregg I: The role of viral infection in asthma and bronchitis. Royal College of Physicians of Edinburgh. Publication No. 46, 1975.
9. May JR: The bacteriology of chronic bronchitis. Lancet, 2:534, 1953.
10. Lees AW, McNaught W: Bacteriology of lower respiratory tract secretions, sputum and upper respiratory tract secretions in 'normals' and chronic bronchitis. Lancet 2:1112, 1959.
11. Laurenzi GA, Potter RT, Kass EH: Bacteriologic flora of the lower respiratory tract. N Engl J Med 265:1273, 1961.
12. Burns MW, May JR: Haemophilus influenzae precipitins in the serum of patients with chronic bronchial disorders. Lancet 1:354, 1967.
13. Miller DL, Jones R: The bacterial flora of the upper respiratory tract and sputum of working men. J Path Bact 87:182, 1964.
14. Tagar I, Speizer FE: Role of infection in chronic bronchitis. N Engl J Med 292:563, 1975.
15. Pines A, Raafat H, Greenfield JSB, Linsell WD, Solari ME: Antibiotic regimens in moderately ill patients with purulent exacerbations of chronic bronchitis. Br J Dis Chest 66:107, 1972.
16. Francis RS, May JR, Spicer CC: Chemotherapy of bronchitis. Br Med J 2:979, 1961.
17. Medical Research Council: Value of chemoprophylaxis and chemotherapy in early chronic bronchitis. Brit Med J 1:1317, 1966.
18. May JR, Delves DM: Ampicillin in the treatment of Haemophilus influenzae infections of the respiratory tract. Thorax 19:298, 1964.
19. Hughes, DTD: Single-blind comparative trial of trimethoprin–sulphamethoxazole and ampicillin in the treatment of exacerbations of chronic bronchitis. Br Med J 4:470, 1969.

156

20. Pines A, Raafat H, Greenfield JSB, Marshall MJ, Solari M: Penicillin, ampicillin and cephaloridine in severe exacerbations of purulent bronchitis in elderly patients. Br J Dis Chest 65:91, 1971.
21. May JR, Ingold A: Amoxycillin in the treatment of chronic non-tuberculous bronchial infections. Br J Dis Chest 66:185, 1972.
22. Willey RF, Gould JC, Grant IWB: A comparison of ampicillin, erythromycin and erythromycin with sulphametopyrazine in the treatment of infective exacerbations of chronic bronchitis. Br J Dis Chest 72:13, 1978.
23. Davies RJ: Skin tests and total IgE levels in reported asthma in middle age. Clin. Allergy 9:417, 1979.
24. Cotes JE: Lung Function Assessment and Application in Medicine, Blackwell Scientific Publications, Oxford, 1975, p. 386.
25. May JR, May DS: Bacteriology of sputum in chronic bronchitis. Tubercle (London), 44:162, 1963.
26. Rawlins GA: Liquefaction of sputum for bacteriological examination. Lancet 2:538, 1953.
27. Rawlins GA: Cytological examination of sputum in relation to its macroscopic purulence. J Clin Path 8:114, 1965.
28. Roberts DE, Cole P: Effect of tobacco and nicotine on growth of Haemophilus influenzae in vitro. J Clin Path 32:728, 1979.
29. Oswald NC, Harold JT, Martin WJ: Clinical patterns of chronic bronchitis. Lancet 2:639, 1953.
30. Reid RD, Fairburn AS: The natural history of chronic bronchitis. Lancet 1:1147, 1958.
31. May JR, Peto R, Tinker CM, Fletcher CM: A study of Haemophilus influenzae precipitins in the serum of working men in relation to smoking habits, bronchial infection and airway obstruction. Am Rev Respir Dis 108:460, 1973.
32. Fletcher CM, Peto R, Speizer FS, Tinker CM: A follow-up study of the natural history of obstructive bronchitis. In: Bronchitis III (Orie NGM, Van der Lende R, eds.), Royal Vangorcum, Assen, 1970, p. 103.
33. Howard P: A long term follow up of respiratory symptoms and ventilatory function in a group of working men. Br J Ind Med 27:326, 1970.
34. Fletcher CM, Peto R, Tinker C, Speizer FE: The Natural History of Chronic Bronchitis and Emphysema. Oxford University Press, Oxford, 1976.

DISCUSSION

Chairman: Thank you very much for this clear introduction. There are many questions. You did a very nice comparative study. Why did you not include a placebo, because you could then compare three drugs with one placebo.

Dr Davies: Most physicians feel, as I did at the start of this study, that antibiotics really have a place in the management of acute exacerbations. I do not want to say they do not have a place, but I am really not as sure as I was. I think we have to qualify the group we are looking at. We are looking here at a group of patients admitted to hospital and it is possible that a nice warm room and good food, physiotherapy and nice nurses make a whole lot of difference to an old man with chronic bronchitis and may make him well as quickly as an antibiotic. I do not know. I think it is time to reassess the role of placebo therapy.

Chairman: I quite agree.

Dr Gould: I, like Dr Davies, have been very interested in this disease for some time. I do not think it is necessarily confined to its definition as 'The English Disease', and his picture of the grossly disabilitated respiratory crumble, as we call them in the North, certainly could be replicated in Edinburgh and Glasgow.

I think there is a difference in taking the types of cases that Dr Davies has taken, as he himself has defined them, as hospital admissions and comparing them with the great bulk of chronic bronchitics with acute exacerbations that others, for example ourselves, have examined, who are seen at home. By definition, I would expect that the ones that are seen in the hospital are much further along the road of debilitating respiratory disease. I think that this is not an infection per se. Probably, infection has no significant part to play in the natural pathogenesis of the disease.

However, talking about acute exacerbations of chronic bronchitis and their treatment in general, I think that in the general population we are rather more correct in considering the treatment of a relatively isolated episode which may recur quite frequently in some people, although not now as frequently as it used to before the Clean Air Act of 1954 and perhaps the reduction in smoking habits, at least in certain sections of the population. Numerous studies that have been carried out in the past show that an antibiotic such as ampicillin has a significant benefit over the placebo. The placebo studies have been carried out and I agree it would be very nice if one could get ethical permission and cooperation to repeat these studies; but a drug like ampicillin or amoxicillin can be taken as a good standard for comparison of other drugs. I was very interested to see that in your study erythromycin and co-trimoxazol are statistically about the same. Another measurement which is different from the domestic case is the fact that you failed to isolate Streptococcus pneumoniae and Haemophilus influenzae from more than 45%. This again is another characteristic of the difference between the type of exacerbation that you were seeing and the type of exacerbation that occurs in the somewhat younger case of acute exacerbation who is working and going about his normal duties. To that extent, I would suggest there is a difference in the type of case that you are assessing and in whom antibiotics do have a role to play in the management of an acute exacerbation. Taking this as an isolated instant it is an infectious disease in its own right, and not related to the natural progression of the basic underlying disease.

Dr Davies: I absolutely agree. There are two very different populations. We have got to be aware of that.

Dr Butzler: I have three short questions. In how many cases did you find an association of pneumococcus and H. influenzae? Second, how did three times 500 mg of erythromycin stearate influence the bacterial flora? Third: in how many cases did you obtain cure of your infection and persistence of bacteria? Dr Gould demonstrated some years ago that you can have a persistence with a cure.

Dr Davies: My answer to the first question is that we had 14 patients from whom we isolated Haemophilus influenzae, in 10 cases Strep. pneumoniae and the majority of these cases coincided.

The second question, that in every case where Streptococcus pneumoniae had been found, initially it was not present in the sputum at the end of the treatment period. Haemophilus influenzae was present in three out of the 14 or 15 from whom we grew it after the treatments. May I mention one other interesting association that we found in our population? This is in part an answer to Dr Gould's question. You will see from our group, after all, we had been getting them to stop smoking cigarettes, and we actually had half our population as ex-cigarette smokers. We found a highly significant correlation between the presence of cigarette smoking and the isolation of Haemophilus influenzae. Indeed, as you are all aware, this is what Robert May showed in the late '60s and early '70s: indeed the presence of Haemophilus influenzae and the antibodies against the H 1 antigen of Haemophilus influenzae is much more common among smokers than among non-smokers, even if they did not have chronic bronchitis.

Recently, a paper by Peter Cole and David Roberts from the Brompton Hospital showed that the presence of tobacco extract or indeed pure nicotine greatly enhances growth of Haemophilus influenzae in culture media. So, one of the explanations for finding rather less Haemophilus in our group than we might otherwise have expected was again: we are seeing a change and it may be different from the out-patient population who may well be going on smoking cigarettes. I think that the proportion of ex-smokers is much higher than in the majority of previous studies.

Dr Dijkman: I notice you have been describing a part of the natural history of chronic bronchitis, as described by Fletcher. In fact, he made a distinction between the hypersecretion syndrome and the obstruction syndrome. And what I saw in your figures perhaps was a mixture of both, but mostly the hypersecretion component.

I would like to ask you whether it would be possible to follow the patients who had infections and study whether antibiotics influence the course of the respiratory function over a period of years.

Dr Davies: Yes. I think this is almost a continuation of the same sort of argument that we are having. As Dr Gould was perhaps suggesting there may well be a group of people who are predominantly mucous-hypersecretors in the population, who have chronic bronchitis by definition. They may get acute episodes of infection in the chest which might be looked at as an acute infectious disease. Perhaps we shall hear more about this group from Professor Grob, which he would see in general practise far more than I would see. There may well be differences between that group and the group we have, who, of course, are the mixed picture of the obstructive and mucous-hypersecretors. There is no doubt that respiratory function can decline without any mucous hypersecretion, as you are well aware. Indeed in Fletcher's studies, up to 25% of the individuals had increased rate of decline of respiratory function without any mucous hypersecretion, and vice-versa. You can have plenty of mucous hypersecretion with no increased rate of decline of lung function. So, there certainly are two groups. I hope I made it quite clear at the onset that we studied a group of old respiratory cripples.

Dr Thompson: What is the contribution of viral infections and mycoplasma infections in causing these exacerbations?

Dr Davies: Looking at the world literature, people would say that only 45% of exacerbations of chronic bronchitis are due to identifiable viral or bacterial infections. Thus viruses play a very important role. This could be perhaps supported by the original long-term administration of antibiotic studies using tetracycline where time of absence from work was reduced by administration of antibiotics but not the number of exacerbations. This suggests that something else non-bacterial was precipitating the exacerbations. I think you can quote almost anybody's figures for the relationships of viruses and exacerbations, depending on which country and the isolation technique. Mycoplasma has certainly been reported, but not to such an extent.

Dr Van der Straeten: You mentioned two exogenous factors in chronic bronchitis: air pollution and cigarette smoking and you deny an allergy factor. There has been a study on local hypersensitivity reaction in chronic bronchitis.

Dr Davies: I know this study by Barry Kay. He found really high levels of IgE and many eosinophils in the sputum. I do not know how you define chronic bronchitis. A lot of people believe there are many eosinophils in the sputum. In our group, we did not find significant numbers, at least not to the extent reported by Barry Kay.

He also found both evidence of histamine and factors which had SRSA-like activity in sputum. This is very interesting. The immunological studies of Robert May show the presence of precipitins and later work shows that these were largely of IgG class. Others produced some evidence to show the presence of IgE antibody against Haemophilus influenzae antigens, again the H-1 antigen. It is an attractive hypothesis to suggest that there is indeed a sort of type I-allergic response against antigens, that might persist in the respiratory tract, following H. influenzae infection. These results need to be repeated.

Dr Van der Straeten: There has been a study of Patterson about sensitizing apes. He aspirated tracheal fluid and injected it into a non-sensitized ape. If he challenged it with the same antigen, he had local reaction.

Dr Gould: I wonder about the etiological agents in the exacerbations, I do not think that there is anything mutually exclusive about viruses being the prime movers in perhaps even the majority of exacerbations, just as in other respiratory-tract infections. I would agree with the figure of 45% as the lower level of virus agents as prime movers in these exacerbations. There is a fair amount of evidence for this.

In relation to the bacteria and the haemophilus story, I think the plot is even thicker. If you restrict yourself to definable Haemophilus influenzae, then the figures may indeed be less than 100%. But, we have tried to show, you can get nearer to 100% if you take into account other haemophilus species. I think that Haemophilus parainfluenzae, Haemophilus hemoliticus and Haemophilus para-hemoliticus may occur in very large numbers in acute exacerbations of chronic bronchitis, similary to Haemophilus influenzae. If you are going to take one as a possible etiological agent, you might as well take the other and treat accordingly.

Chairman: Dr Kunst, could you please summarize our findings in chronic bronchitis?

Dr Kunst: In the out-patient clinic of the Departments of Infectious Diseases and Pulmonary Diseases of the University Hospital in Leiden, called the 'Bronchitis polikliniek', a study was performed on the effectiveness of amoxicillin versus ampicillin for the treatment of chronic bronchitis. The criteria for admission to the trial were the same as those mentioned by Dr Davies, except that the exacerbations were not so severe as to require admission. Over a period of 28 months 38 patients were studied, 22 of them for the whole period and 16 for a mean period of 19 months (range 4–26 months). The respiratory function of 28 of these patients is summarized in Table 1.

The arrangement was that when the patient had an exacerbation he or she

Table 1. Exacerbation of obstructive respiratory diseases.

Forced expiratory volume (% VC)	Number of patients
≤ 34	
35-44	12
45-54	6
55-64	3
≥ 65	5

would come to the out-patient clinic and deliver early-morning sputum. On that occasion a physical examination was performed as well as macroscopic evaluation of the sputum. These investigations were repeated at 14-day intervals and the washed sputum was coloured by Gram stain and cultured according to Mulder's method on days 1, 14, 28 and 42. When the investigator considered the patient's physical problems as an acute exacerbation of the chronic bronchitis, antibiotic treatment was assigned single-blind at random: amoxicillin 750 mg three times daily or ampicillin 1000 mg four times daily, for a period of 10 days. Registration included the number of days of increased sputum produced before and after antibiotic treatment, days of greenish-yellow sputum after treatment, increased dyspnoea, bed-rest, fever, and increased malaise.

Although in the year preceding the trial all 38 patients had chronic mucopurulent bronchitis with at least two exacerbations requiring treatment, only 65 exacerbations were registered in 25 patients during the 28-months period. Thirteen patients had no exacerbations at all. Furthermore, 53 'anamnestic' exacerbations in 24 patients, 26 of whom were treated with various antibiotics, were not

Table 2: Sputum Findings (University Hospital, Leiden).

	Before treatment	day 14	day 28	day 42
No. of sputa examined macroscopically	65	50	39	19
Mucous	22	23	21	14
Purulent	43	27	18	5
Culture:				
No growth	11	11	3	0
H. influenza	14 ⎫	3 ⎫	6 ⎫	2 ⎫
Strep. pneumoniae	11 ⎬ 29	4 ⎬ 9	4 ⎬ 12	1 ⎬ 5
B. catarrhalis	11 ⎭	2 ⎭	2 ⎭	2 ⎭
Others	3	7	3	0

registered. Most of these patients had not come to the clinic because they were too sick. In 27 patients with exacerbations who did not receive antibiotic treatment, spontaneous recovery occurred. During the course of an acute exacerbation the purulence and the amount of sputum both diminished (Table 2), as did the number of H. influenzae, Strep. pneumoniae, and B. catarrhalis.

Regular (bimonthly) examination of 180 random sputum samples of the patients showed that 65 were mucous and 115 mucopurulent (22 showed no growth, 54 H. influenzae and/or Strep. pneumoniae and/or B. catarrhalis, and 39 other bacteria). In about 45% of the mucopurulent sputa, however, these bacteria were regularly found in follow-up cultures, and might therefore be considered inhabitants of the bronchial tree.

No significant difference between amoxicillin and ampicillin was found as to the course of the acute exacerbations in chronic bronchitis. However, there was a negative correlation between the number of days with increased sputum production before and after antibiotic treatment, which supports Davies's conclusion, that either antibiotic treatment is not very helpful or the role played by bacterial infection is insignificant.

Furthermore, we did not find any important role played by viral infection, either. Viral serology was performed during 42 exacerbations, and no significant rise in antibodies against respiratory viruses of Mycoplasma pneumoniae was found.

The most important side effect observed for the antibiotic was diarrhoea, mostly related to ampicillin (31% of treatment episodes versus 5% in amoxyllin).

13. TREATMENT OF RESPIRATORY INFECTIONS IN CHILDREN

K.F. KERREBIJN

INTRODUCTION

In the antibiotic treatment of respiratory infections in children it should be known if the infections are recurrent* or incidental, and if they are of viral or bacterial origin. Recurrent respiratory infections generally occur in children with decreased local or general defence mechanisms because of a predisposing disease, such as an asthmatic predisposition, cystic fibrosis, immune deficiency or congenital anatomical malformation.

As has been observed in children with cystic fibrosis, recurrent inflammations of the pulmonary airways in childhood may result in anatomical changes which can be compared with those observed in chronic bronchitis in adults, i.e. hypertrophy of the mucous glands, increase in the number of goblet cells, origin of these cells further towards the peripheral bronchi, and thickening of the basal membrane between mucosa and submucosa.

It is assumed that this is caused by bacterial infections [1, 2]. Although so far any such observations in children with recurrent bronchial infection on a 'constitutional' basis are lacking, it can be assumed that similar anatomical changes may also occur in some of these patients.

It is likely that they are not always completely reversible, because in a great number of children X-ray abnormalities such as line shadows or mottling continue to exist even after bronchial infections have been stopped for a considerable time.

Irreversible deformations may also occur, although this is comparatively rare. Examples are: Pneumatoceles after a pneumonia due to Staph. aureus, the Mc. Leod-Swyer James syndrome due to destruction of the pulmonary vessels and damage to the airways and bronchiectasis due to an inadequately treated pneumonia. Epidemiological studies also indicate that respiratory infections at a young age predispose to cough and sputum production in later periods of life. An example is the study of Colley et al. [3] who found a strong correlation between recurrent cough below the age of 2 and at the age of 20.

---

* Recurrent bronchial infections occur frequently, and run a longer course than usual.

It is often difficult to determine to what extent recurrent respiratory symptoms in children are caused by infections or by other factors. This is particularly so in patients with an asthmatic constitution who belong to the above mentioned category with an increased risk of only partly reversible changes. In particular, 'low-grade' bronchial infections, where the defense mechanisms insufficiently hamper the infectious process, may cause symptoms of long duration without temperature, leucocytosis or an increased sedimentation rate.

The differentiation between a viral and a bacterial infection cannot be made on clinical grounds. Neither does the blood picture give a decisive answer: with a bacterial infection leucocytosis may be missing; with a viral infection a moderate leucocytosis may exist. Only the examination of sputum combined with the effect of antibiotic treatment confirms that a bacterial infection had been responsible for the symptoms. In young children especially, it may be difficult to obtain an adequate sputum sample for culture and microscopy on leucocytes and bacteria.

ANTIMICROBIAL THERAPY

In respiratory symptoms in which sputum cannot be obtained, it is often difficult to decide if an antibiotic treatment is indicated, and whether a narrow-spectrum antibiotic will do or if a broad-spectrum antibiotic is necessary. Our present policy is as follows: we advise general practitioners in cases of incidental infections of the respiratory tract for which they consider treatment with antibiotics to be indicated, to start with a narrow-spectrum penicillin. If there is no improvement within 48 h and the presence of a bacterial infection is still probable, the treatment should be changed to a broad-spectrum antibiotic which attacks not only grampositive microorganisms but H. influenzae as well. We use amoxycillin, cotrimoxazole or, depending on the age of the patient, doxycycline.

Doxycycline is given at any age if we suspect a Mycoplasma pneumoniae infection because of the epidemiology or a positive cold agglutination. In patients with recurrent infections of the airways, H. influenzae is so often involved that we consider a treatment with a broad-spectrum antibiotic from the very onset as preferable. This also holds if sputum cultures are not available. In the case of cystic fibrosis the choice of antibiotics should be determined by the bronchial flora, but in small children this is not always possible. In that case we start with flucloxacillin and make a further choice according to the clinical response. To allow an immediate start of the treatment we give the antibiotic in stock to those patients in whom we know that antibiotic treatment gives a rapid improvement. They start according to previous experiences, which are obtained by recording the effect of the treatment in previous periods of illnes after starting at variable moments (prodrome, first symptoms, fully developed symptoms).

The daily or prophylactic use of antibiotics in patients with recurrent or chronic infections of the airways is avoided if possible to prevent the colonisation of the bronchi with gram-negative microorganisms, such as Pseudomonas aeruginosa. This may, but does not always, occur in patients with a decreased local defence against infections due to anatomical deformations or mucous plugs as in cystic fibrosis or bronchiectasis, or in patients with immunodeficiencies or other disturbances in the general defense mechanisms.

Daily antibiotic treatment, however, cannot always be avoided. We give antibiotics daily if this appears to be the only way to keep cough and sputum production within acceptable limits. In children with severe chronic respiratory disease, studies on the effect of daily antibiotics versus antibiotic treatment at infectious periods on the frequency and duration of the infectious exacerbations and the long-term course of the disease are lacking. The choice of daily treatment antibiotics depends on the bronchial flora; in the case of H. influenzae we use amoxycillin, cotrimoxazole or doxycycline and with staphylococci preferably flucloxacillin. A recent study of Williams [4] shows that trimethoprim induces thymine-dependency in Staph. aureus, which creates difficulties in the isolation and identification of the bacteria. This should be kept in mind when patients with a chronic infection with staphylococci, as in cystic fibrosis, are treated with cotrimoxazole.

The daily treatment of Pseudomonas aeruginosa infections is difficult to perform. The only effective way in which pseudomonas can be attacked is by means of i.v. administration of tobramycin (or gentamicin) combined with ticarcillin (or carbenicillin or azlocillin). This provides no problems for short-term periods of treatment, but is almost impossible if it has to be continued for more than 3–4 weeks. Aerosols with aminoglycosides may be used, but it is questionable if the small concentrations attained in the pulmonary airways when the antibiotic is delivered by aerosol have more than a marginal effect. Moreover, the more severe the airway obstruction, the less effectively will the aerosol be deposited.

A recent study by Wood et al. [5] indicates that the concurrent administration by aerosol of ethylenediaminetetra acetate (EDTA) which retards the growth of most pseudomonas strains in vitro and systemic or aerosolized antibiotics, may result in a more effective treatment of pseudomonas infections.

The oral treatment of pseudomonas infections is not effective. However in practice we often give cotrimoxazole or doxycycline daily to our cystic fibrosis patients who are severely colonised with pseudomonas. Although these antibiotics should not be effective we have the impression that they often contribute in keeping the equilibrium in these patients. Clinical side effects from daily antibiotic treatment are rare. In CF patients the original microorganism is often, but not always, replaced by Pseudomonas aeruginosa.

In the treatment of pneumonia the following points should be borne in mind: (1) Though bacterial pneumonias in children are most frequently caused by

Table 1. Antibiotic treatment of respiratory infections*.

Name	Route of administration	Dose in mg/kg/24 hr		Number of gifts per 24 hr
		Age < 9 weeks	Age ⩾ 9 weeks	
Aminoglycosides				
streptomycin	i.m.	10–25	20–40	2
kanamycin	i.m./i.v.	15	15	2
gentamicin	i.m./i.v.	5–10	5–10	3
tobramycin	i.m./i.v.	3–10	3–10	3
amikacin	i.m./i.v.	10 initial dose 7.5 thereafter	10–15	2
Cephalosporins				
cephalexin	oral	50–100	25–100 (max. 4 g)	3–4
cephalotin	i.m./i.v.	20–30	40–60	4
cephamandol	i.m./i.v.		50–150	4
Chlooramphenicol	oral		50–100	4
	i.m./i.v.	15–25	50–100 (max. 3 g)	2
Cotrimoxazol	oral		6 trimethoprim	
	i.v.		30 sulphame- thoxazol	2–3
Lincomycines				
lincomycin	oral		30–60	3–4
	i.m./infusion	10–20	10–20	2–3
clindamycin	oral		10–30	3–4
	i.m./infusion		20–40	3–4
Macrolides				
erythromycin	oral	25–40	30–50	4
– ethylsuccinate				
+ stearate	oral	20–30	30–50	4
– ethylsuccinate	i.m.	20–30	20–30	3
– lactobionate	infusion	10–20	30–50	3–4
Penicillins				
Na benzylpenicillin	i.m./i.v.	30 000–50 000 IU/kg	30 000–50 000 IU/kg	4
phenoxymethylpenicillin				
phenoxyethylpenicillin phenoxypropylpenicillin azidocillin	oral		30	3
ampicillin	oral	50	50–100	3–4
	i.m./i.v.	50–100	50–200	3–4
amoxicillin	oral	25– 75	25– 75	3–4
	i.m./i.v.	50–250	50–250	3–4
pivampicillin	oral		25– 75	3–4

Table 1 (continued)

Name	Route of administration	Dose in mg/kg/24 hr		Number of gifts per 24 hr
		Age < 9 weeks	Age ⩾ 9 weeks	
pivmecillinam	oral		20– 60	4
meticillin	i.m./i.v.	50–100	100–200	4–6
cloxacillin	oral	50	50–100	3–4
	i.m./i.v.	25– 50	50–100	4–6
flucloxacillin	oral	50	50–100	3–4
	i.m./i.v.	25– 50	50–100	4–6
carbenicillin	i.m./i.v.	50–200**	50–200**	4–6
		250–1000***	250–1000***	
ticarcillin	i.m.	50–100	50–100	4
	i.v.	200–300	200–300	4
Rifamycins				
rifampicin	oral	20	20	2–3
			(max. 600 mg/ 24 hr)	(1 in tuber- culosis)
Tetracyclines				
doxycycline	oral/i.v.		4 first day 2–4 next days	1(–2)
minocycline	oral/i.v.		4 first day 2–4 next days	1(–2)
rolitetracycline	i.m./i.v.		10	1–2
tetracyclinechloride	oral		20–40	3
Tuberculostatic agents				
ethambutol	oral	10	10 (max. 300 mg/ 24 hr)	1
isoniazide	oral	10	10–15 (max. 500 mg/24 hr)	1
protionamide	oral		15–20	1
pyrazinamide	oral		20	1
rifampicin	oral	20	20 (max. 600 mg/24 hr)	1
streptomycin	i.m.	10–20	15–20	1

* From Lubsen et al., 1980 [6].
** Coli and proteus infections.
*** Pseudomonas infections.

pneumococci, the possibility of other bacteria must always be considered. The most common are: H. influenzae (especially in neonates and patients with recurrent bronchial infections), and Staphylococcus aureus (especially in infants; in patients who are in close contact with staphylococci and during influenza epidemics). Of the non-bacterial pneumonias, that caused by Mycoplasma pneumoniae is most frequently encountered. Its clinical symptoms often resemble the symptoms of influenza.

(2) In immunocompromised patients inflammations caused by Pneumocystis carinii or cytomegalovirus may occur.

Pneumococcal and *streptococcal* infections are always treated with a narrow-spectrum penicillin.

Haemophilus influenzae infections are treated with amoxycillin, cotrimoxazole or doxycycline.

Staphylococcal infections are treated with penicillin and with (flu)cloxacillin in the case of β lactamase forming bacteria. If the staphylococci are also insensitive to (flu)cloxacillin, another antibiotic is used depending on the sensitivity. Infections with Pseudomonas aeruginosa are treated with high doses of carbenicillin or ticarcillin combined with an aminoglycoside, preferably tobramycin. Because of the varying sensitivity of pseudomonas for carbenicillin, a quantitative determination of the sensitivity is advocated. In the few cases who were insensitive to tobramycin we have used amikacin.

Klebsiella pneumoniae and Enterobacter infections are treated with either the combination of ticarcillin with gentamicin or a cephalosporin (i.e. cephamandol) with kanamycin.

In a severely ill patient it is of vital importance to start immediately with the right antibiotics. Until the results of the bacteriological examination are known, we usually start in such cases with the combination of cephamandol with kanamycin, to which almost all microorganisms are sensitive with the exception of Pseudomonas aeruginosa.

For dose recommendations in children, the reader is referred to our publication 'Antibacteriële geneesmiddelen in de algemene praktijk' [6]. The recommended dose for children is summarized Table 1.

REFERENCES

1. Reid L, DeHaller R: Lung changes in cystic fibrosis. In: Cystic Fibrosis: a symposium (Hubble HV ed.), Chest and Heart Association, London, 1964, p. 21.
2. Esterly JR, Oppenheimer EH: Cystic fibrosis of the pancreas: structural changes in peripheral airways. Thorax 23:670, 1968.
3. Colley JRT, Douglas JWB, Reid DD: Respiratory disease in young adults: influence of early childhood lower respiratory tract illness, social class, air pollution, smoking. Br Med J 3:195, 1973.

4. Williams RF: Persistence of thymine-dependent staphylococci in CF sputum: a pitfall for the unwary. In: Perspectives in cystic fibrosis: Proceedings of the VIIIth. International Congress on Cystic Fibrosis (Sturgess JM, ed.), Canadian Cystic Fibrosis Foundation, Toronto, 1980, p. 352.
5. Wood RE, Klinger JD, Thomassen MJ, Cash HA: The effect of EDTA and antibiotics on Pseudomonas aerogunosa isolated from cystic fibrosis patients: a new chemotherapeutic approach. In: Perspectives in cystic fibrosis: Proceedings of the VIIIth. International Congress on Cystic Fibrosis, (Sturgess JM, ed.), Canadian Cystic Fibrosis Foundation, Toronto, 1980, p. 365.
6. Lubsen N, Kerrebijn KF, Waardhuizen JP van (eds.): Antibacteriële geneesmiddelen in de algemene praktijk. Beecham Research Laboratoria, Amsterdam, 1980.

DISCUSSION

Dr. Van der Meer: I would like to know from Dr Kerrebijn what he thinks is the purpose of performing bacteriological investigations in patients with recurrent respiratory infections, especially in children.

Dr Kerrebijn: I think it is important to try to investigate whether a bacterial infection is present and if we decide that a bacterial infection plays a role in causing the symptoms–and we decide this on the grounds of a positive culture plus leukocytes in the sputum–then we treat with antibiotics, on the assumption that recurrent bacterial infections may damage the bronchial wall. But this is not at all proven, as I have indicated.

Dr Davies: I am very interested in this particular point. While I agree that the point is not proven by any means, there are long-term prospective studies in schoolchildren from both England and America looking at the relationship between decline in peak expiratory flow rate and episodes of infection. This does not seem to be–at least according to a study in schoolchildren in Kent that comes up with the finding that those children who had recurrent episodes of infection had lower peak expiratory flow rates. So, it is a very important issue. Whether the infection started all off or whether there is an underlying factor which makes them more liable to infection in this first instance is what we would all like to know.

Dr Kerrebijn: This is true. There are a number of studies now. One is a prospective study and two or three are retrospective studies which indicate, indeed, that respiratory symptoms at a very young age are presumably caused by infections; but this is very difficult to prove. As I have indicated, this may increase the risk of respiratory disease at young adult ages. I think this has been well established now. We do not know exactly the role of infection.

We know that often at one or two years of age the symptoms can be treated very adequately and quickly with antibiotics, and we assume that bacterial infections then often play a role.

Dr Simon: How do you define the term 'recurrent respiratory-tract infection' in children? What is 'normal' in children: Another questions is: what is the frequency in children, depending on the age, with which primary viral infection changes into a secondary bacterial infection?

Do you believe that in younger children, i.e., in the first or second year of life, the risk is higher and the use of an antibiotic should be advocated?

Dr Kerrebijn: The definition is not very clear-cut. We define recurrent infections as those which are more frequent and run a longer course than usual, but I am well aware of the fact that this is not sharply delineated. I think you will agree that the child who has, say, five to six episodes per year, with coughing and sputum production which lasts, without treatment, for more than a week, has more symptoms than the average child. But, there is a gradual difference between normal and clearly abnormal. There is no real cut-off point. Others defined respiratory infections as those recurring twice a year, not very frequently at any rate. My feeling was that perhaps this frequency could be within the normal range, but even with such a low frequency, these children had a high risk of having respiratory symptoms at the age of twenty.

In answering your second question, I think that it starts mostly with viral infections at some time¦ but I do not know how often a bacterial component follows. But, again, the differentiation between a purely viral infection and mixed infection is very difficult to perceive. There are several studies in the literature in which people have tried to prove viral infection, but the findings of more than 50%–by means of culture and increase of titres–are very rare.

Dr Hilvering: Is it not as important to treat these patients for their bronchospasm as for their bacterial infection? I have the impression that if you do not do this, the infection often lasts longer and it can become more severe than when you use antibiotics alone. This applies to grown-up patients as well as children.

Dr Kerrebijn: If there is a clear spasm, I agree. But this is not always the case in young children; sometimes they wheeze a little. If you give a spasmolytic drug and you try to get objective data–and this can only be done by means of the stethoscope–it often turns out that no clear bronchospasm exists together with the infection. In some patients it does, but in many it does not. In these patients, a spasmolytic drug is not indicated.

Dr Hilvering: In my opinion, in adult patients, there often is a bronchospasm.

Dr Kerrebijn: I fully agree. In older children it is so. As soon as they have reached school age, there is often a spasm. But I am referring to the ages of one to three years.

Dr Hilvering: Then I would even suggest to you to give corticosteroids, because this is a very good spasmolytic drug in cases of coexisting asthma.

Chairman: I agree, but we should use corticosteroids as little as possible. When you have other means, as a chest physician, I would think you would try to use them first.

Dr Van Boven: I was just thinking about the ironic situation in which we find ourselves: we have a full range of antibiotics available and the more we have the greater are our doubts concerning their use. My question is: is it possible to find some circumstantial or indirect evidence regarding the use of antibiotics in these respiratory diseases of early childhood? In late childhood and early adulthood–as compared with 20 or 30 yr ago–we see much less serious pathological lesions of the lungs. Is there no difference when you compare the results with those of 10, 20 or 30 years ago?

Dr Kerrebijn: I think it is quite obvious that in, say children in modern countries bronchiectatic disease has nearly disappeared. What we see of bronchiectasis is in children from Turkey, Morocco, the Cape Verde Islands, etc. Why? Histories have not been obtained from these children. They were probably not treated with antibiotics when they had viral pneumonia. But there are strong indications that frequent use of antibiotics, in any case, diminishes largely anatomical deformations.

Chairman: Do you not think that vaccination against pertussis and measles as well as immediate antibiotic treatment of serious bacterial complications during a viral infection–for example, following influenza–will have fostered a better condition for the lungs, in Western countries.

Dr Kerrebijn: Oh, yes. Without any doubt, vaccination against whooping cough has contributed largely to the decrease in bronchiectasis. Vaccination against measles probably will, but in this country routine measles vaccination has been performed only in the last five years and not on all children. But, the decline was seen in the '60s.

Chairman: Even so, after measles, when they had some bacterial complications, these children were treated adequately.

Dr Van der Waaij: Continuous prophylactic treatment we avoid as much as we can. Only on few occasions we feel that there may be an indication to treat a child continuously for some time; and it is in fact only to permit the original flora to return after a prolonged treatment with a penicillin-like antibiotic, which, as I indicated yesterday, also interferes with the oropharyngeal flora. In fact it is only in cystic fibrosis patients, that we sometimes do that, but not for patients with chronic lung infections, which is your largest group of patients.

Dr Kerrebijn: I completely agree with this. We nearly never treat patients with recurrent pulmonary infections without gross anatomical deformations with continuous antibiotics. We only do that in patients with cystic fibrosis, not routinely–we try to avoid it as much as possible. Sometimes, on clinical grounds, it is unavoidable, such as in patients with bronchiectasis, until they undergo surgery, if this is possible. But, this is a very small group and I have not much experience in this.

Dr Van der Waaij: In that case, do you monitor the oropharyngeal flora?

Dr Kerrebijn: No, we do not.

Dr Davies: One thing that particularly interests me is whether people here generally use antibiotics in the treatment of asthma? This is a widespread habit in England, still. I am in the middle of a study trying to look at the benefits of antibiotics or otherwise, in asthma. But, I can tell that 50% of exacerbations of asthma in England are treated with wide-spectrum antibiotics. This is in adults, but it really relates to your sort of problem. I know your problem of wheezy bronchitis. What is your common practise? What do you advise for asthmatic episodes in children?

Dr Kerrebijn: I advise in clear-cut asthma cases not to give antibiotics. But, as I indicated, it is sometimes not very easy to differentiate between an infection and an asthmatic component together with the infection. Our practice then in the hospital is as follows: we try to obtain sputum and this is easier with in-patients than with out-patients; then we examine the sputum for leukocytes and eosinophils. If there are many eosinophils, we do not give antibiotics even if we see bacteria in the gram preparation. If there are no eosinophils and many leukocytes in the sputum plus bacteria, we give antibiotics. Sometimes, we see that the antibiotics help, but sometimes they do nothing.

Dr Davies: The interpretation of leukocytes in sputum is difficult, is it not? But I am fascinated by that. It is exactly what we are doing in adults in a controlled trial of antibiotics versus placebos in adult episodes of asthmatics and looking at daily sputum cytology.

In what proportion of your asthmatic children do you think, from your studies on sputum, that bacteria have a part to play?

Dr Kerrebijn: I would say we give antibiotics to these children in about 30% of the cases, not more. It depends a littele on the age. If they are young children, aged one to three years–they are not real asthmatics by then, but they have wheezy bronchitis–we give antibiotics much more frequently than we would give to the age group five to six years old.

Chairman: Dr Kerrebijn do you have information about the role of the slow maturation of the secretory IgA system in the pathogenesis of chronic bronchitis? And a question for Dr Davies. When these children become adults, does secretory IgA play a role?

Dr Kerrebijn: I think there are many papers on secretory IgA and bronchitis, also in children. We have not done any work on that. I think we should differentiate between IgA in the blood and IgA in the sputum. A low IgA content in the blood is fairly rare and is not correlated with recurrent respiratory symptoms, as has been shown by a number of people. There have been a number of papers on low IgA in sputum correlated with recurrent respiratory infection. But I am not familiar enough with the technique which enables you to know whether the procedures followed in these studies were always adequate or not. The outcomes are controversial.

Dr Davies: I would be happy to comment about that. I think that the point you raised last is actually very important. Once again, there are great technical problems in how IgA is being measured, making the results in the literature rather conflicting. There are a number of studies, but there have been technical problems involved in many of them, which does not allow us at this time to make a clear-cut decision as to whether people with bronchitis either get it because they have deficient secretory IgA, or indeed whether the disease itself eventually leads to production of low secretory IgA. Here are two possibilities and I do not know really convincing evidence that has been backed up by several investigators to answer these questions.

Chairman: That was also my impression from the literature.

Dr Van der Waaij: We have found, to our surprise, not so much decreased IgA secretion in these patients, but many of them do fail to produce secretory component. It is amazing, but quite a number of those patients have a decrease in or absence of secretory component, particularly in the saliva and the nasal secretions. The meaning of this is yet unknown.

Chairman: I have never seen any patient with that.

14. ANTIBIOTIC TREATMENT OF MYCOPLASMA PNEUMONIAE INFECTIONS

M. VAN DER STRAETEN

Although various mycoplasma species have been found in man, only Mycoplasma pneumoniae is known to affect the human respiratory tract. Together with viral infections, Mycoplasma pneumoniae is among the most common causes of acute infections of the upper and lower respiratory tract in man [1]. Of a large series of patients with clinical symptoms of an acute respiratory infection seen over a 10-year period (1965–1975), 23% had antibody titers indicating the possibility of a recent infection with Mycoplasma pneumoniae.

Infections are seen in all age groups, but the incidence is highest among

Table 1. Clinical manifestations of Mycoplasma pneumoniae infection.

Upper respiratory tract
  Pharyngitis
  Otitis media, bullous myringitis

Lower respiratory tract
  Tracheobronchitis
  Exacerbation of airway obstruction
    in patients with COLD
  Pneumonia
    Acinar pattern
    Interstitial pattern
  Hilar adenopathy
  Pleural effusion
  Lung abscess
  Adult respiratory distress syndrome
  Diffuse interstitial fibrosis

Extrapulmonary complications
  Hematologic
  Neurologic
  Cardiovascular
  Dermatological
  Gastrointestinal
  Musculoarticular

children and young adults, especially those living in closed communities, e.g. college students or military recruits, because the disease is transmitted as a droplet infection. Most of the outbreaks occur at the end of the autumn or in the beginning of the winter and are closely associated with epidemic influenza. During such outbreaks, 10 to 20% of all cases of acute respiratory infection are due to Mycoplasma pneumoniae. In contrast to the influenza outbreaks, M. pneumoniae tend to be spread over a period of several months. Epidemics seem to occur every four to five years. Between the outbreaks Mycoplasma infections are known to be endemic with a low incidence of about 2%.

The clinical picture presented by M. pneumoniae infections range from an inapparent infection to severe multilobular pneumoniae and a number of bizarre non-respiratory complications. It is rare for M. pneumoniae infections to be fatal, but a number of well-documented case reports have been published [2–4]. Upper respiratory tract symptoms and bronchitis are the most common manifestations and the clinical course of these infections is usually mild and self-limiting.

Table 2. Mycoplasma pneumoniae infection: non-respiratory manifestations.

Blood
  Hemolytic anemia
  Diffuse intravascular coagulation
  Thrombocytopenic purpura
  Paroxysmal cold hemoglobinuria

Cardiovascular system
  Myocarditis
  Pericarditis
  Raynaud's syndrome

Central nervous system
  Meningo encephalitis
  Ascending paralysis (Guillain-Barre)
  Transverse myelitis
  Cranial nerve palsies

Skin
  Exanthemas: macular, vesicular, bullous, petechial
  Erythema multiforme: minor and major (Stevens-Johnson syndrome)

Gastro-intestinal tract
  Hepatitis
  Pancreatitis

Musculoskeletal system
  Polyarthralgia, myalgia
  Acute arthritis

Between 3 and 10% of the patients infected with M. pneumoniae subsequently develop pneumonia [5, 6].

The incidence of pneumonia due to this organism is low but is probably underestimated because chest radiograms are seldom made in cases of seemingly mild respiratory tract disease and patients with extensive pulmonary infiltrates may show minimal clinical signs of pneumonia on physical examination , which not arrows a suspicion of pneumonia (Table 1).

A number of extrapulmonary complications have been described, such as myocarditis, acute hemolysis, meningomyelitis, and erythema multiforme, which are among the most frequent systemic complications (Table 2), but the rates at which pulmonary and systemic complications occur in M. pneumoniae infections is not known. It has been argued that M. pneumoniae should be suspected in patients presenting with otherwise unexplained acute neurological symptoms [7], and the same holds for other systemic complications of M. pneumoniae infections, such as acute perimyocarditis and the Stevens-Johnson syndrome. There is increasing evidence that host factors play a significant role as determinants of disease expression. Pneumonia due to M. pneumoniae may be at least partially an immunopathologic process. This possibility must be taken in account in the treatment of these M. pneumoniae infections. The foregoing explains the need for antibiotic treatment of M. pneumoniae infections. Until the identification of the 'Eaton agent' (PPLO organism) treatment of 'the primary atypical pneumonia' was a controversial subject. As could be expected from their lack of rigid cell wall, mycoplasmas are resistant to antibiotics whose antibacterial action is at that site, e.g., penicillins and cephalosporins. Definitive information on the efficacy of an antibiotic in M. pneumoniae infections was provided by the controlled study of Kinston et al. [8] with dimethylchlortetracycline. Their double blind clinical trial on the efficacy of a moderate oral dose of this drug (900 mg daily for 6 days) showed that the duration of clinical symptoms (coughing, fatigue, malaise, fever) was significantly reduced, progression of the pneumonia was arrested, and clearing of the infiltrates was accelerated.

Although most patients recover without treatment, chemotherapy reduces the duration of clinical symptoms even though it may not eradicate the organism [9–11]. Jao and Finland [12] assessed the in vitro susceptibility of 5 strains of M. pneumoniae to 21 commonly used antibiotics. Erythromycin was clearly the most active of the agents tested, M. pneumoniae being sensitive to minute amounts. Among the tetracycline group, dimethylchlortetracycline appeared to be the most active. However, all the tetracyclines are active in inhibiting growth of M. pneumoniae at concentrations easily attained in the serum of patients given adequate and well-tolerated doses. The minimum inhibiting and mycoplasmacidal concentration in microgram per ml was determined for the 21 antibiotics and plotted against the activity (Figs. 1 and 2).

The general sequence of decreasing activities is essentially the same for the

178

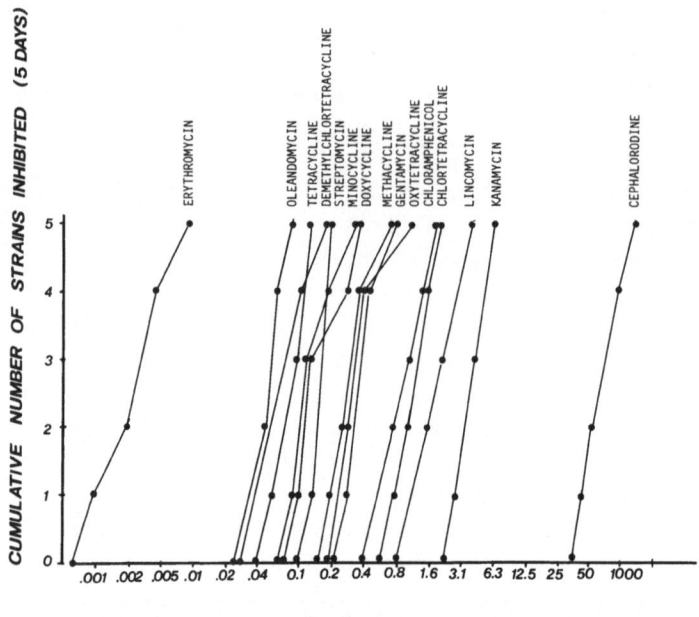

Fig. 1. Minimum inhibiting concentration for 5 strains or M. pneumoniae [12].

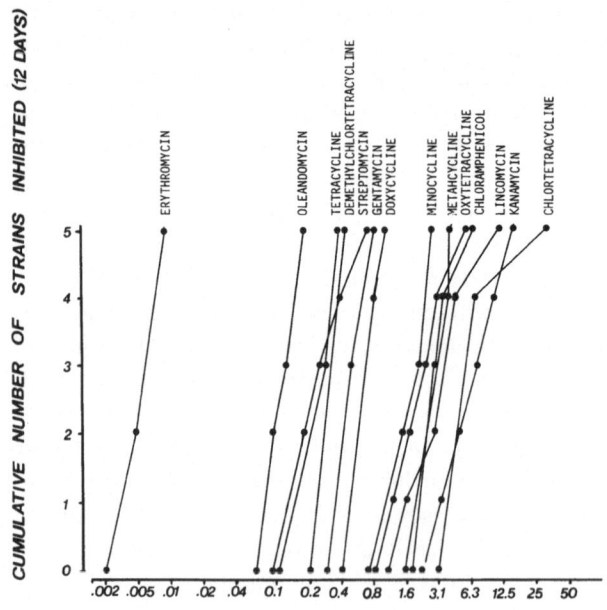

Fig. 2. Minimum mycoplasmacidal concentration for 5 strains of M. pneumoniae 12].

inhibitory and cidal effects. The difference between MIC and MPC is greatest for the tetracyclines. Gentamicine is also active against M. pneumoniae in vitro, and has been found to be beneficial in vivo [13,14].

Slotkin et al. [15] also studied the effect of antibiotics on M. pneumoniae in vitro and reached the same conclusions as Jao and Finland [12]. Both tetracycline and erythromicin were found to inhibit growth of M. pneumoniae in a static fashion in vitro although erythromycin was 50 times more effective on a weight basis. Shames et al. [16] assessed the clinical response to 6 antibiotics (demeclocycline hydrochloride, erythromycin stearate, erythromycin ethylsuccinate, tetracycline hydrochloride, methacycline hydrochloride, and trileandomycin) in the treatment of M. pneumoniae infections as compared with a control group. The aim of this study was to answer two questions: (1) do antibiotics alter the clinical course of M. pneumoniae infections, and (2) are any of the proposed antibiotics significantly more effective than the others? Concerning the first question they concluded that the morbidity of patients infected with M. pneumonia was reduced; the treated groups fared significantly better than those serving as controls. Various indications of improvement were also assessed: the duration of fever, the duration of hospitalization, the time required for clearing of lung infiltrates, and the persistence of the organism during and after therapy. Definite conclusions as to the relative effectiveness of the different antibiotics could not be drawn. With both erythromycin and the tetracyclines, the clinical response was rapid but the organism often persisted in the sputum.

The effectiveness of these 6 antibiotics in eradicating M. pneumoniae from the respiratory tract could not be determined in their study either. Erythromycin and tetracylines are generally considered the drugs of choice for the treatment of M. pneumoniae infections. Adults are usually given erythromycin stearate 1–2 g or doxycycline 200–100 mg per day. In children erythromycin ethylsuccinate 50 mg per kg per day is successful and well tolerated [17]. These antibiotics give rapid control of fever, usually within 2 days, and rapid relief of the cough associated with improvement of the general condition. After this successful clinical treatment the carrier state can persist for several weeks. In experimentally induced infections, too, there is a continous shedding of M. pneumoniae during and after treatment with tetracycline and erythromycin.

Wenzel et al. [18] reported similar salutary effects obtained in M. pneumoniae infections with a new macrolide, josamycin. There was no significant difference in the number of hospital days or fever days between the erythromycin group or the josamycin group. Both antibiotics were well tolerated at doses of 2 g daily per os. Blood levels were more than 100 times higher than the reported median mycoplasmacidal drug levels for josamycin (0.015 μg/ml) and erythromycin (0.007 μg/ml).

Although clindamycin is active as a static agent in vitro, at least one controlled blind trial failed to demonstrate its efficacy in man [19].

From the work of Hers [20] it can be concluded that superinfections associated with M. pneumoniae infection are uncommon and usually occur in children. The most prevalent secondary bacterial pathogens are Haemophilus influenzae, Streptococcus pneumoniae, and Staphyloccus aureus. Mixed infections with respiratory viruses such as influenza and adenovirus, have also been described on the basis of rising complement fixation titers.

Grady and Gilfillan [21] reported that 22 out of 27 cases of Legionnaires' disease showed seroreactivity against M. pneumoniae, this coincidence of seroreactivity is unexplained. The authors suggest that M. pneumoniae-positive cases should be investigated with respect to Legionnaires' disease. The findings of cross-reacting antibodies to Micropolyspora faeni in patients infected with M. pneumoniae [22] led to the suggestion that the surface membranes of these species may carry a common polysaccharide antigen. Elevated complement-fixation titers for M. pneumoniae reported in patients with sarcoidosis suggest an aspecific titer rise in patients with a strong immune response.

Fernald et al. [23], who studied the pulmonary immune response to M. pneumoniae infection, put forward a hypothesis concerning the immune pathogenesis of mycoplasma disease. Two properties of M. pneumoniae seem to correlate extremely well with its pathogeniticity in man:

(1) It is a non-invasive organism characterized by adherence to the mucosa of the tracheobronchial tree, behaving like a surface parasite of the respiratory epithelial cells although it has occasionally been found in the pleura, middle ear, and cutaneous bullae. M. pneumoniae has never been detected intracellularly.

(2) It produces hydrogen peroxide and other toxic substances presumed to be responsible for the damage to the epithelium. The H_2O_2 is thought to be responsible for hemolysis as well as for the stimulation of cold agglutinins. It has been suggested that H_2O_2 alters the antigenicity of the antigen on the red cell, thereby inducing the elaboration of anti-I antibodies of the IgM class which can agglutinate erythrocytes in the cold. The pathogenesis of systemic manifestations such as hemolytic anemia, central nervous system syndromes, and skin manifestations, is unknown. The involvement of extrapulmonary systems may be related to antigenic simularities between parasite and host. Antibodies of the IgM class, reactive with other tissues, are detectable by a complement fixation method utilising normal brain, heart, lung, or liver as antigen. These antibodies against heart, lung, and brain may mediate the auto-immune phenomena occasionally seen in seriously sick patients.

Recent studies on experimental infections with M. pneumoniae in hamsters have underscored the importance of cell-mediated immunity. Biberfeld et al. [24] and Fernald [25] demonstrated lymphocyte transformation in patients infected with M. pneumoniae, and Arai et al. [26] and Biberfeld [27] described a cell-mediated immune response to M. pneumoniae in experimentally infected hamsters detected with the macrophage migration inhibition test. Brunner et al. [28] obtained similar results in experiments with guinae pigs. These authors suggest the possibility of a sensibilization to M. pneumoniae in man after an infection during early childhood. In these sensitive patients the stimulation of lymphocytes could liberate toxic substances damaging the tissues.

Taylor [29] reported experimental infection with Mycoplasma pulmonis in mice. Normal mice infected with M. pulmonis often show marked peribronchial and perivascular infiltration by lymphocytes, whereas in thymectomized irradiated mice the lesions were much less prominent.

Foy et al. [13] reported 4 cases of M. pneumoniae infection in patients with immune deficiency. Chest X-rays of these patients suggested that pulmonary infiltrates may be the result of an immunological reaction. Accumulation of lymphocytes and plasmocytes accounts for the peribronchial infiltration characterizing the radiological appearance of pneumonia. Patients with immune-deficiency syndromes, who lack a normal B-cell production, are seriously sick without radiological evidence of pneumonia. All these studies suggest that the host cellular immune response may be deleterious in cases of severe infection. The cell-mediated cytotoxicity could be the cause of respiratory failure. On the basis of this concept of immunopathology in M. pneumoniae infection, seriously sick patients have been treated with corticosteroids in combination with first-choice antibiotics. Noriega et al. [30] have described a case of fulminant interstitial pneumonitis due to M. pneumoniae in which steroid therapy was beneficial. In other cases the combination of antibiotics and corticosteroid therapy was unsuccessful, perhaps because treatment was started late and severe irreversible tissue damage was present [3, 4].

Although erythromycin and tetracyclines are the drugs of choice for the treatment of M. pneumoniae infections, resistance to these antibiotics has been mentioned by several authors. Niitu et al. [31] were the first to report the isolation of a M. pneumoniae strain highly resistant to erythromycin and resistance has been reported for tetracyclines [32]. Stopler et al. [33] published results on sensitivity to antibiotics tested in 28 Mycoplasma strains. In patients treated with erythromycin or tetracyclines, the sensitivity of the micro-organism to these drugs remains at the same level after 8 days of treatment. This findings suggested that there are certain sites in the nasopharynx where the antibiotic fails to reach a satisfactory level, and therefore even a sensitive M. pneumoniae strain may persist on the mucosa. The minimum plasmacidal concentration were determined. Nine isolates gave rise to mutants resistant to high concentrations of erythromycin. None of the strains produced mutants resistant to tetracycline.

182

SUMMARY

Recent reviews of respiratory-tract affections caused by M. pneumoniae underscore the benign and often subclinical course of the infection. Severe pneumonia with a reticular or acinar pattern is certainly unusual and a fatal outcome is rare, but the incidence of both is underestimated. Erythromycin and tetracyclines are the first-choice antibiotics. There is evidence indicating the importance of immunopathogenic mechanism in provoking pneumonia and even respiratory failure.

REFERENCES

1. Krech U, Price PC, Jung M: The laboratory diagnosis and epidemiology of mycoplasma pneumoniae in Switzerland. Infection 4:33, 1976.
2. Fischman RA, Marschall KE, Kislak JW: Adult respiratory distress syndrome caused by mycoplasma pneumoniae. Chest 74:471, 1978.
3. Reigner Ph, Domenighetti G, Feihl F, Bonjour JPh, Perret Cl: Syndrome de détresse respiratoire aigu sur infection à mycoplasme. Sch Med W 110:220, 1980.
4. Kaufman JM, Cuvelier CA, Van der Straeten M: Mycoplasma pneumonia with fulminant evolution into diffuse interstitial fibrosis. Thorax 35:140, 1980.
5. Murray HW, Masur H, Senterfit L, Roberts R: The protean manifestations of mycoplasma pneumoniae infection in adults. Am J Med 58:229, 1975.
6. Levine DP, Lerner AM: The clinical spectrum of Mycoplasma pneumoniae infections. Med Clin N Am 62:961, 1978.
7. Twomey JA, Espir ML: Neurological manifestations and Mycoplasma pneumoniae infection. BMJ 2:832, 1979.
8. Kingston JR, Chankock RM, Mufson MA, Hellman LP, James WD, Fox HH, Mankoma C, Boyers J: Eaton agent pneumonia. JAMA 176:118, 1961.
9. Grayston JT, Alexander ER, Kenny GG, Clarke ER, Fremont JC, macColl Wa: Mycoplasma pneumoniae infections. J Am Med Assoc 191:369, 1965.
10. Maisel JC, Babbitt LH, John TJ: Fatal Mycoplasma pneumoniae infection with isolation of organisms from lung. JAMA 202:287, 1967.
11. Jones MC: Mycoplasma pneumoniae. Practioner 203, 1969.
12. Jao RL, Finland M: Susceptibility of Mycoplasma pneumoniae to 21 antibiotics in vitro. Am J med Sc 253:639, 1967.
13. Foy HM, Ochs H, Davis SD, Kenny GE, Luce RR, Mycoplasma pneumoniae infections in patients with immunodeficiency syndromes: report of four cases. J Infect Dis 127:388, 1973.
14. De Vos M, Van der Straeten M, Druyts E: Myocarditis and severe bilateral bronchopneumoniae caused by Mycoplasma pneumoniae. Infection 1:60, 1976.
15. Slotkin R, Clyde W, Denny FW: The effect of antibiotics on Mycoplasma pneumoniae in vitro and in vivo. Am J Epid 86:225, 1967.
16. Shames JM, George RB, Holliday WB, Rash JR, Mogabgab WJ: Comparison of antibiotics in the treatment of Mycoplasma pneumoniae. Arch Int Med 125:680, 1970.
17. Ruhrmann G, Holthusen W, Noack B: Klinische und röntgenologische Beobachtungen zum Verlauf von Infektionen mit Mykoplasma pneumoniae im Kindesalter. Infection 5:47, 1977.
18. Wenzel RP, Graven RB, Davies JA, Hendley JO, Hamory BH, Gwaltney JM, Jr: Protective efficacy of an inactivated Mycoplasma pneumoniae vaccine. J Infec Dis 136:S204, 1977.

19. Smilack JD, Burgin WW, Moore WL, Sanford JP: Mycoplasma pneumoniae pneumonia and clindamycin therapy. JAMA 228:729, 1974.

20. Hers JP: Clinical aspects of infection with Mycoplasma pneumoniae. Proc roy Soc Med 61:1325, 1968.

21. Grady GF, Gilfillan RF: Relation of mycoplasma pneumoniae seroreactivity, Immunosuppression, and Chronic Disease to Legionnaires' Disease. Annals Int Med 90:607, 1979.

22. Davies BH, Edwards JH, Seaton A: Cross-reacting antibodies to micropolyspora faeni in mycoplasma pneumoniae infection. Clin Allergy 5:217, 1975.

23. Fernald GW, Clyde WA, Denny FW: Nature of the immune response to mycoplasma pneumoniae. J Immunol 98:1028, 1967.

24. Biberfeld P, Sterner G: Cell mediated immune response following mycoplasma pneumoniae infection in man. Clin exp imm 29:17, 1974.

25. Fernald G: In vitro response of human lymphocytes to mycoplasma pneumoniae Inf Imun 5:552, 1972.

26. Arai S, Hinuma Y, Matsomoto K, Nakamura T: Delayed hypersensitivity in hamsters infected with Mycoplasma pneumoniae as revealed by macrophage migration inhibition test. J Microbiol 15:509, 1971.

27. Biberfeld G: Macrophage migration inhibition in response to experimental Mycoplasma pneumoniae infection in the hamster. J Immunol 110:1146, 1973.

28. Brunner H, James WD, Horswood RL, Chanock RM: Experimental Mycoplasma pneumoniae infection of young guinea pigs. J Infect Dis 127:315, 1973.

29. Taylor G: Immunity to mycoplasma infections of the respiratory tract: a review. Roy Soc Med 72:520, 1979.

30 Noriega ED, Simberkoff MS, Gilroy FJ, Rahal JJ: Life-threatening Mycoplasma pneumoniae. JAMA 229:1471, 1974.

31. Niitu Y, Hasegawa S, Suetake T, Kubota H, Komatsu S, Horikawa M: Resistance of mycoplasma pneumoniae to erythromycin and other antibiotics. J Pedr 76:438, 1970.

32. Jones GR, Borthwick RC: Mycoplasma pneumoniae resistant to oxytetracycline: two case reports. Brit J Dis Chest 67:119, 1973.

33. Stopler T, Gerichter CB, Branski D: Antibiotic-resistant mutants òf Mycoplasma pneumoniae. Isr J Med Sc 16:169, 1980.

DISCUSSION

Dr. Mattie: If a clinical trial comparing any drug, not especially antibiotics, does not show any differences among these drugs, the most probable explanation is that none of them have any effect. Is there any indication that antibiotic treatment is better than no treatment at all?

Dr Van der Straeten: There are differences between the control and treated groups. These were real differences; in days of fever, in days in hospital. But there were no differences between erythromycin and the tetracycline group.

Dr Stam: Is it still true that cold agglutinins in the blood are present in a higher percentage in this disease?

Dr Van der Straeten: Yes. But other causes, such as viral infections can also give increased titres.

Chairman: But in what other viral infections? As far as my experience goes, it is very rare and more recent figures indicate that about 50% of mycoplasma infections have an increase in cold agglutinins. It might be dependent upont the epidemic strain. But I am not aware of any of the other viral infections that give, so consistently, an increase of cold agglutinins.

Dr Van der Straeten: In the epidemic at the end of 1975, we had 26 cases of Mycoplasma pneumoniae infections and we have seen cold agglutinins in one-fourth of those cases.

Dr Dijkman: I have investigated about 50 patients with other viral infections in connection with this, and only in 2% of the cases cold agglutinins were present.

Chairman: So, when there is an increase in cold agglutinins, it could be a mycoplasma infection.

Dr Kunst: It may be that cold agglutinins are negative in the first week and may become positive in the second week. This may perhaps explain the low incidence when the test is not repeated after some time. In Van der Straeten's series, when was this done?

Dr Van der Straeten: It was done in the second, and third or fourth week; it was done twice.

Dr Butzler: To come back to Dr Mattie's question. There is no doubt that antibiotics are effective in mycoplasma infection. There was a very nice study published about two years ago in a Scottish medical journal. Here, there was even a difference between erythromycin and the tetracyclines. However in our experience, there is no difference between doxycycline and erythromycin. Can you confirm this?

Dr Van der Straeten: Yes, I think in the majority of our cases we have treated, we are using doxycyline and only in a small percentage erythromycin; but there was no difference. The fever disappeared within 24 to 48 hr.

Dr Gould: Dr Butzler refers to work in general practice and this was carried out over two separate years. So it was not controlled. If I remember correctly, there was no difference between the measurable clinical effects of tetracycline versus erythromycin. There was a clinical impression that erythromycin was more effective in reducing the number of relapses, relapses were a feature of this series.

Dr Meenhorst: The findings you quoted where they found cross-reactivity between mycoplasma and legionella antibodies have not been substantiated by other investigators. However, we have found, in a couple of patients, also a rise against mycoplasma and legionella and this may well be a double infection, but no proof exists that this is really a fact. Also cross-reactivity with bacteroides fragilis has been found. This is the beginning of the story, let us see what will come out of the serology in the future.

Dr Van der Meer: A number of patients who come into the hospital with severe pneumonia will receive antibiotic therapy in which aminoglycocides will be included. Is there some activity of the aminoglycosides for mycoplasma? Do you think this has any importance for the therapy?

Dr Van der Straeten: No. I just mentioned it because it was based on the wrong diagnosis of septicemia–the patient had myocarditis–in which ampicillin and gentamycin were given. I think the first choice in treatment is certainly tetracycline.

Dr Van der Waaij: My question concerns the antibodies. I noticed that you have cross-reacting antigens not only in legionnaires' disease, but also for several tissue antigens. Are they responsible for the remote side effects in the skin, the brain and the pericardium? And if so, how long do these antibodies persist? Do they require separate treatment?

Dr Van der Straeten: I think that the nature of immunity in mycoplasma infections is largely unknown. I do not think anybody knows exactly how long they persist.

Dr Van der Waaij: But do they have no pathological significance; are they not responsible for the skin rash?

Dr Van der Straeten: Perhaps. There is a possibility.

Dr Dijkman: You raised the point of treatment with corticosteroids in addition to treatment with antibiotics. This probably had something to do with the ideas about the immune phenomena. Is there any evidence for this?

Dr Van der Straeten: There are only case reports on treatment with corticosteroids. We had one patient who died after several weeks because we had started too late. The tissue damage was so severe, as discovered at the autopsy, that I believe corticosteroids could have been beneficial if they had been given early enough.

Chairman: What you have demonstrated with your patient is that it has no harmful effects. When you have, for example, a patient suffering from very severe hemolytic anemia—which can occur—or just other serious complications, in these cases you apparently can give corticosteroids in large doses for a short period and might just put a stop to the complications. But, in general I would not advocate treating mycoplasma infections with antibiotics and corticosteroids, because then we are on the wrong track.

Dr Van der Straeten: I agree with you, but that patient came in a severe hypoxemia and at that point, we must give corticosteroids.

Chairman: That still does not prove that it is an immunological phenomenon.

Dr Van der Straeten: No, it is hypothetical.

Chairman: It might just be used to stop the normal inflammation, without any immunological etiology.

Dr Kerrebijn: In practice mycoplasma infections pose difficulties in diagnosis, not only in children but also in adults. My question is: if a mycoplasma infection is treated from the very beginning with either doxycycline or erythromycin, can we then expect a rise in antibody titre and cold agglutinins? When a child comes in and we suspect a mycoplasma infection, then we mostly start with one of these drugs. Later, we get results from the lab and these are frequently negative and we do not know whether this has been a mycoplasma infection or not. This is a difficulty which I cannot solve. What is your suggestion?

Dr Van der Straeten: I think that in this case, there is no mycoplasma involved. The incubation time is about two weeks. I suspect that in all the cases of mycoplasma infection the complement fixation test was positive after two weeks.

Dr Kerrebijn: So, you think that if we do not have a rise in antibody titre we do not have a mycoplasma infection.

Dr Van der Straeten: Yes, but I have no irrefutable arguments.

Dr Van der Meer: I thought the incidence of Mycoplasma pneumoniae infections in chuldren below the age of six was very low. Is this not true?

Dr Van der Straeten: It is possible. I cannot say exactly.

Dr Kunst: Dr Van der Straeten, do you know the recurrence rate? How is the immunity status after having had a Mycoplasma pneumoniae infection? The person who has had mycoplasma, how long does he stay immune and does he run a chance of getting again a mycoplasma infection with another type in one, two years.

Dr Van der Straeten: I think this is difficult to answer. A second infection certainly exists, it has been suggested that those cases suffering from pneumonia must have been infected in childhood a first time, and there will be some sensibilization; the second infection will produce pneumonia. However I cannot say how long the duration of the immune status will remain.

Dr Kunst: You know of no references dealing with persons who have a recurrent mycoplasma infection?

Dr Van der Straeten: I have not seen any references regarding this subject.

Chairman: Do you know anything about different antigenic strains, because these could probably explain situations where you have a new infection. Do they exist? Apparently, there is no knowlegde of this.

Dr Thompson: Has any attempt been made to prepare a vaccine?

Dr Van der Straeten: This has been done, yes. There is a formalized vaccine, with an effect of about 40%.

Chairman: But it is not in use in humans?

Dr Van der Straeten: No.

Chairman: I suppose it is a veterinarian vaccine.

15. DEVELOPMENTS IN ANTIBIOTIC TREATMENT OF RESPIRATORY INFECTIONS IN GENERAL PRACTICE TOWARDS BETTER PRESCRIBING

P.R. GROB

INTRODUCTION

A general practitioner in the United Kingdom may well see, especially in the busy winter months, up to three hundred patients per week, of these three quarters could easily be suffering from respiratory disease. These will largely consist of the elderly or children. Much of this respiratory disease will be seen in a very early phase, and may well be self limiting and self-curing. In addition many of the conditions will have a viral rather than a bacterial aetiology. This then presents the major difficulty for the general practitioner, whom do you select for treatment with antibiotics, taking into account the size of the problem and the early nature of the disease as it is encountered in family practice.

Another difficulty is engendered by the constraints of time and cost. It is probably more cost effective to give every patient complaining of a sore throat a course of penicillin rather than send off a pathological specimen to the laboratory, wait for the return of the bacterial culture and then provide those patients with a bacteriologically proven sore throat, with the appropriate antibiotic treatment. The first course of action, in providing everybody with an antibiotic may be good economics but it is clearly unsatisfactory medicine.

A further problem has been engendered by the attitude of many patients in expecting the physician to provide an instant 'cure' for every trivial complaint. The 'pill for every ill' syndrome. Fortunately in recent years there appears to be a swing away from this artificial expectation. The general public now seem to be recognizing that every action has its risk and its benefit.

Another problem is encountered by the fact that it is sometimes harder to do nothing than embark on a course of activity. This may produce anxiety on the part of the patient and his medical attendant, and it also may result in the general practitioner missing the early manifestations of a severe and treatable condition.

It is much easier for our hospital colleagues to adopt a policy of observation when the patient is under constant medical or nursing surveillance, and any deterioration can be rapidly monitored.

Another dimension was added to the aspects of prescribing by the work of Howie [1] who showed in a number of elegant experiments, that doctors were influenced by a great many other factors other than their scientific observations, as to whether they prescribed or not. For example, students who had an exam to take the next week were more likely to be given an antibiotic than a student who was going on his vacation. A mother with a young family, who was obviously harassed, would be more likely to receive an antibiotic, as would someone who lived a long way from the consulting doctor. The word 'science' means knowledge, the difficulty with the art of medicine lies in the application of that knowledge.

The family doctor being a generalist, also has difficulties in assessing the rival claim and merits of newer antibiotics. He is subjected to a battery of advertising from the pharmaceutical industry, some of which may by very conflicting in its nature. Sometimes he has difficulty in sorting in his own mind what are real therapeutic advances in any field of chemotherapy, and what are only spurious claims.

Although it may appear the general practitioner has considerable difficulties in the field of prescribing, he also has advantages over his hospital colleagues. As he works in the community, he is able to detect any departure from the normal patterns of disease.

It has been found that a fruitful way of tackling this problem has been to instal a small incubator with some culture plates in the practice itself. When the general practitioner detects the presence of a new epidemic of respiratory disease in the community he is able to take bacterial and virological swabs for analysis from the first dozen or so patients who present with this new condition. Although the results are back in a few days in the case of bacterial infection, he is then able to base the treatment of the subsequent patients he sees with more scientific accuracy. Fig. 1 illustrates the small incubator in the author's practice which has served this purpose for many years.

Another method of improving the general practitioner's awareness of the spectrum of the disease which he is encountering in his community has been developed by the use of sentinel practices in the United Kingdom. The Epidemic Observation Unit of the Royal College of General Practitioners has about 120 sentinel practices throughout the United Kingdom who make a regular weekly return on a selected number of diseases that they encounter in their practices. These diseases are then analysed centrally and correlated with the data which are available from hospital laboratory sources. Furthermore, some groups of these sentinel practices not only record their epidemiological data, but also underpin this by the appropriate microbiological investigations. Recording doctors are then sent a fortnightly feedback sheet which gives an analysis of the latest patterns of infectious diseases the likely organisms and their sensitivity patterns [2].

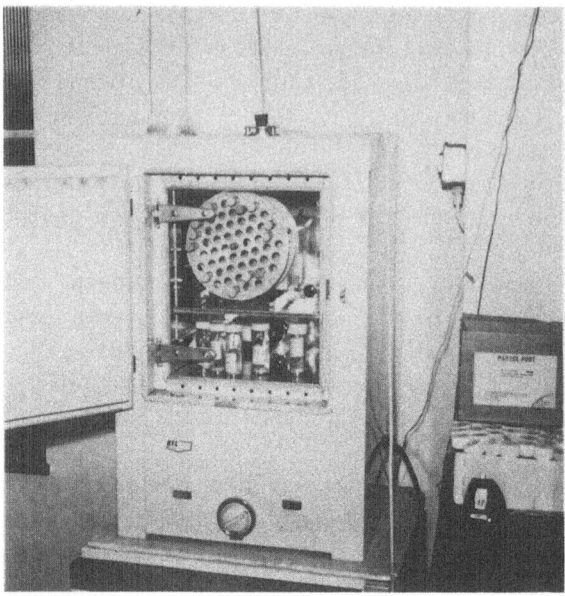

Fig. 1.

PERSONAL PRESCRIBING

The policy pursued by the author with regard to antibiotic prescribing, is to have a limited and small number of drugs which are well used, their documentation and contra-indications known, and they are usually older cheaper antibiotics.

The following are the quidelines which have helped the author in his treatment of different respiratory infections in general practice.

Sore Throat

It is difficult to distinguish clinically the appearance of a sore throat caused by bacterial infection from that of a viral infection. A clue made be given however that pain seems to be a predominant feature of sore throats caused by the maemolytic streptococcus. This organism accounted for 42% of sore throats in the author's practice, but these occurred in two large waves usually in the spring and in the autumn.

If it was apparent that the patient is suffering from a bacterial sore throat, oral penicillin was one of the drugs of first choice, however, a problem arises by the fact that about 14% of patients in the author's practice were allegedly allergic to this drug. The word 'Allegedly' is used advisedly, as not infrequently patients say that they or their children are allergic to penicillin and the evidence for this is not very substantial. What they usually mean is that the child vomited or was upset by the administration of the drug rather than a true allergic reaction. However, the general practitioner would be ill advised to give penicillin to a patient who had told him that he was allergic to this drug. There is no good and satisfactory test which will distinguish true allergy in this drug. As a second choice therefore erythromycin has been found to be extremely satisfactory in the treatment of bacterial sore throats.

Otitis Media
This is one of the commoner presenting symptoms in young children, and in this condition where greater tissue penetration is required ampicillin, or more recently, amoxycillin has been the drug of first choice.

Respiratory Disease in Children
It is sometimes very difficult to isolate the pathogenic organism in respiratory disease in children, and in this case general practitioners and paediatricians find that a wider approach may in fact be useful. Erythromycin and amoxycillin seem to be tolerably well accepted in this particular condition when the aetiology remains uncertain. The only major problem with erythromycin, seems to be that about 10% of children find it causes abdominal colic and vomiting.

Whooping Cough
Experience in the recent outbreak of whooping cough in the United Kingdom, has suggested that erythromycin, given for fourteen days is probably the drug of choice in this condition. The earlier the drug is administered the more effective it seems to be, conversely if the child has been coughing for two or three weeks any antibiotic appears to be relatively ineffective in modifying the course of the illness.

Lower Respiratory Tract Infection in Adults
This usually present as exacerbations of chronic bronchitis. In these patients amoxycillin, trimethaprim or tetracyclines all seem to be of value. These drugs may also be of value if given prophylactically during an outbreak of epidemic influenza to the vulnerable elderly bronchitic. In the last large outbreak of influenza in the United Kingdom in 1968, two waves of death were encountered. Firstly, there seemed to be a number of patients who died within 12 hr of contracting the disease, and clearly any antibiotic treatment would probably be of little value in these patients; there was a second wave of deaths which occurred two to three weeks after the initial illness, in which a post mortem examination showed signs of pneumonic changes. In this latter group it is believed that the administration of antibiotic may well be of benefit.

The use of antibiotics in prophylaxis in chronic bronchitis has been of considerable interest to general practitioners [3]. A four year study untertaken by the author on a group of chronic bronchitics who were given erythromycin 500 mg t.d.s. for the first ten days of each of the six winter months, suggests that there may be a place for this type of therapy in the treatment of this disease, the predominating infecting organism was H. influenzae or Strep. pneumonae. Chronic bronchitics who have pursued this regime for four years seem enthusiastic about it and objectively have recorded fewer outbreaks of acute exacerbation of bronchitis as the years progress. It is however too early to say whether this pattern of improvement will be permanent as time advances.

A very fruitful way of encouraging general practitioners to adopt better prescribing habits has been developed in the United Kingdom by the use of audit groups. In these audit groups small numbers of doctors, record their prescribing habits and these are then analysed centrally by the Birmingham Research Unit of The Royal College of General Practitioners and subsequently they receive their prescribing patterns for discussion and analysis. This has proved to be a remarkably effective way of altering doctors prescribing habits [4, 5]. These audit groups are further improved if a cost analysis of their prescribing can also be provided [6].

It is considered important that when the balance of prescribing versus non prescribing is assessed, that it is recognized that it is difficult to evaluate the disease that is prevented. Paediatricians fortunately now rarely see children with chronic bronchiectasis, and ear-nose and throat specialists seldom see children with chronic mastoid disease.

In conclusion, it is suggested that the family practitioners have difficulties but also advantages in the field of sensible antibiotic prescribing. It is often stated that general practitioners over prescribe antibiotics and in some cases this may well be true, but it is suggested that there are ways of trying to base prescribing on a scientific footing in general practice.

REFERENCES

1. Howie JGR: Clinical Judgment and Antibiotic use in general practice. Brit Med J 2:1061, 1976.
2. Continuous Morbidity Observations 1975–78. Report from the Epidemic Observation Unit of The Royal College of General Practitioners. Published by The Royal College of Gen. Pract.
3. Grob PR: The use of erythromycin in a general practice. Scottish Med J 22:405, 1977.
4. Crombie D: Practice activity analysis. J Roy Coll Gen Pract 27:265, 1977.
5. Fleming D, Maes RMJ: Facets of practice in the United Kingdom and Belgium. Allgemeinmedizin 1:5, 1980.
6. Singer GE: Department of Health & Social Security, London, Personal communication.

DISCUSSION

Chairman: What is the state of vaccination against whooping cough?

Prof. Grob: It is very low indeed in some areas, especially in the low social class in Wales where it is as low as 7%. Overall, it is about 30% throughout the United Kingdom. We are sorry about this, but this is what happens when people become emotionally involved in this subject. It is almost entirely due to a very vociferous professor–I will not mention who he is–who never ceases to appear on the media, saying how dangerous whooping-cough vaccination is. Not only are the levels of whooping-cough vaccination dangerously low, but there is a knock-on effect that things like poliomyelitic, diphtheria, measles, tetanus, about which we are moderately happy, are falling off as well. It is a very worrying problem when people use emotion instead of logic, causing immense harm.

Chairman: It is an unfortunate situation which, fortunately, is not applicable to The Netherlands. Here, about 94% are vaccinated against all these antigens.

Dr Gould: I wanted to endorse all of what Professor Grob has said and to emphasize, from the point of view of a clinical microbiologist who has a great deal to do with general practitioner services, that the great pool of infection exists more and more in the general non-hospital population. Furthermore, most of the infections that occur in general practice present as the primary disease of the individual, while more and more frequently, when we see infections in hospital they are incidental to the situation of the patient.

Therefore, we come back to this whole question of the access by general practitioners to laboratory services and communication. Apart from those general practitioners who can be encouraged to carry out a little of their own work–as Professor Grob has illustrated–I think what he refers to as communication is most important. However, I would ask him how he can get this information on syndromes occurring within his area quickly enough to be of value? At the present time, you might say that there is a good service from Collindale, C.D.S.C. which gives a great deal of valuable information, but this comes out in some cases rather too late in acute situations. For example, we have had a very sharp outbreak of pneumococcal disease–at least in our geographical area–over the last weeks. This can, at present, only be individually communicated to practitioners verbally or by letter from the laboratories.

I wonder if Professor Grob has any ideas about such situations can be communicated rapidly to be more effective in narrowing down the diseases that require to be watched for and therefore treated, in the community.

Prof. Grob: Of course, the point you make is a salient one. What we are trying to do is to organize the analysis of the regional data at regional level, so that the fortnightly sheet is peculiar to a particular region and is produced by the local regional community physician, who is well aware of what is happening. The pilot study we are doing in the Oxford region seems to be working very successfully indeed. Here, hospital data is being correlated–returns from public-health laboratories–with the sort of pictures and syndromes the general practitioners themselves are reporting. So, it is a marrying-up together of the data which is sent into my unit–which is essentially epidemiological–with the appropriate under-pinning microbiology. I think this is a very fruitful and exciting area for the future.

Dr Thompson: Do you think that in the process of decision making by the general practitioner, there is a place for computer-based programs and do you know if anything in that sphere is being developped in Great Britain?

Prof. Grob: We are very interested in micro-computing in general practice. I would think that your brain is a very much better, facile, subtler machine; it programs into all the variables that I showed on the screen, like, who is your father, what do you want to be doing next week, how far do you live away from the practice. Computers are quite good when you feed them in hard data in science, but I think we have seen that there is more intuition and more skill which are very difficult to program. It is a nice area to look at for the future.

Chairman: Which antibiotics do you use, and how long, how much, and when?

Prof. Grob: You are asking for the secret of the universe in 20 min? Penicillin is my first-line drug–penicillin V by mouth–for respiratory-tract infections, and sore throats. The problem there is that about 14% of my patients are alledgedly allergic to penicillin, in as much as in their records someone has written 'allergic to penicillin'. There is no good test to find if they are, apart from giving it to them which is perhaps not the wisest thing to do. If I am concerned, then erythromycin which I rank as an equal first. For upper respiratory-tract infection in children, I would perhaps prescribe amoxicillin or erythromycin. For Otitis media, I would prescribe either ampicillin or amoxycillin. For lower respiratory-tract infection, I would give the broad-spectrum antibiotics. Unless I am suspecting a mycoplas-ma pneumoniae infection, erythromycin would be my drug of choice.

Chairman: How long do you give antibiotics? There is always disagreement. Do you prescribe for 5, 7, 10 or 14 days?

Prof. Grob: Multiples of five days, because the medicine comes in a five-day bottle. Generally, I find that five days is satisfactory.

Dr Van der Meer: There are a number of new, potent, orally-administerable antibiotics coming on the market nowadays and Dutch general practitioners are first and large-scale users of these drugs. I would like to have your opinion on how we should influence their prescribing habits.

Prof. Grob: This situation is most dangerous. It really touches on the pressures that are brought to bear on general practitioners who work in relative isolation. The spectrum of the diseases we see does not change. As far as I am concerned, I have some quite good remedies–at least they work in my hands–for the commonly encountered diseases. I am very leery about changing my prescribing habits.

How you persuade your colleagues not to be the first to prescribe the latest super 'me too' product is a very difficult problem. We see this as much as you do. I say stick to old, tried, cost-effective remedies. But I may be terribly old-fashioned in this.

Dr Van der Meer: Do you think that the oral cephalosporins, especially the new ones like cefaclor, have a place in general practice?

Prof. Grob: Not in my practice. I could be wrong. I do not know the answers.

Dr Van Boven: I think the points Professor Grob raised were very interesting. The challenge of the future lies in general practice. There is a statement: the general practitioner over-prescribes, over-uses antibiotics. Is this correct?

Prof. Grob: It sounds like saying: when did you last stop beating your wife? It depends. But it deserves a better answer than that. Like generalizations, on some occasions these are true, but I do not think you can generalize like that. It is a very well rehearsed argument; I do not think it is a very well researched argument.

Dr Van Boven: Did the audit show these pictures to general practitioners and did that influence their prescriptions?

Prof. Grob: Yes. Audit is a very powerful way of changing people's behaviour. It must be done in a supportive, non-punitive, non-threatening way. They have done it with Belgian doctors as well as United Kingdom doctors. It is very simple: what do you do? what do I do? how do we differ? That is when the discussion takes place.

We were doing some work with some Belgian doctors. Eighteen of us were prescribing things like ampicillin or what have you for certain conditions. One of us was prescribing chloramphenicol. He quickly recognized that he was out of step. We did not have to say anything.

Chairman: Our behaviour may also change after this meeting.

16. ANTIBIOTIC TREATMENT DURING INFLUENZA VIRUS INFECTIONS

J.H. DIJKMAN

Normally, the defense against invasion of the lower airways by micro-organisms is very well organized. Each breath we take probably brings bacteria into the lower airways. These bacteria either come from the outside air and are not caught by the filtering system of the nose or–more usually–they are added to the inhaled air and thus contaminate during contact with the membranes of the upper airways; these membranes are not sterile and can easily convey commensal flora.

AIRWAY INFECTIONS

Infections of the airways can be classified according to the pathogenetic properties of the micro-organisms involved and the general and local defense mechanisms.

Primary Infections

Here, micro-organisms have pathogenetic properties enabling them to overcome the normal defense (Table 1). Examples are viruses attacking previ-

Table 1. Primary infections of the airways by pathogenic micro-organisms.

Micro-organisms
  Viruses
  Mycobacterium tubercilosis
  Fungi (Histoplasma capsulatum)
  Bacteria (B. pertussis)

Clinical characteristics
  Healthy individuals
  Micro-organisms not present before infection
  Everyone susceptible to infection
  Incubation period
  Outcome depends on specific immunological reaction

ously unexposed individuals who lack specific humoral antibodies. The same holds for some mycobacteria (M. tuberculosis) and Mycoplasma pneumoniae, but only for a few fungi (Coccidioides immitens, Blastomyces spp.) and bacteria (B. pertussis, P. pestis). Bacteria and also fungi commonly found in the upper airways (H. influenzae, Strep. pneumoniae, B. catarrhalis and A. fumigatus and C. albicans, respectively) do not behave like primary pathogens and are only able to produce infections when the normal defense of the host is disturbed.

The host defense system can be disturbed in two ways: immunologically and locally.

Opportunistic Infections

The term opportunistic infection (Table 2) is reserved for infections by micro-organisms occurring in the compromised host, which means the absence of an adequate reaction of the immunological system (production of humoral antibodies or cell-mediated immunity by macrophages and lymphocytes). Such a general disturbance of the defense against micro-organisms may be due to

Table 2. Opportunistic infections of the airways.

Micro-organisms
    Viruses: Cytomegalo virus, Herpes virus
    Protozoa: Pneumocystis carinii
    Fungi: Asperpillus fumigatus, Candida albicans
    Bacteria: Gram-negative species

Clinical characteristics
    Decreased immunological defense
    Micro-organism present before infection
    No incubation period
    Localizations in other organs as well
    Outcome depends on severity and duration of immunological defect

Table 3. Conditional infections of the airways.

Micro-organisms
    Bacteria: H. influenzae, Strep. pneumoniae, B. catarrhalis
    Fungi: Aspergillus fumigatus

Clinical charateristics
    Normal immunological defense
    Micro-organism present before infection
    Recurrent infections
    Local defense disturbed (defective drainage)

diseases impairing the function of these cells or defects established iatrogenically, usually by treatment with cytostatic or immunosuppressive drugs.

The vast majority of patients with bacterial respiratory infections do not, however, have any recognizable immunological defect, but belong to the group of patients with chronic obstructive respiratory disease. Therefore, a third category of airway infections has been introduced, i.e., conditional infections [1].

Conditional Infections

Under this term is understood alterations in the local defense mechanisms in the absence of immunological disorders (Table 3). Examples are an abnormal ciliary function (Kartagener's syndrome, heavy smokers), abnormal mucus production (cystic fibrosis), hypersecretion of mucous, edema, and bronchospasm (chronic

Table 4. Decreased local defense mechanisms.

Bronchial deformation
  bronchiectasis
  tumor
  foreign body

Damaged epithelium
  virus infection
  chemical irritation

Cardiac congestion

Bronchial obstruction (drainage disturbed)
  glandular hypertrophy ⎫
  hypersecretion   ⎬ chronic obstructive lung disease
  oedema bronchial wall ⎪
  bronchospasm   ⎭

Table 5.

	Respiratory infections		
	Primary	Opportunistic	Conditional
Micro-organism			
highly pathogenic	+	−	−
Decreased immunological			
defense	−	+	−
Decreased local defense	−	−	+

bronchitis, asthma, chronic obstructive lung diseases), local changes such as bronchiectasis and obstruction by a tumor or foreign body that impairs normal bronchial clearing. Another example is local destruction of epithelium, providing a porte d'entrée for bacteria, as in viral infections, notably influenza. A list of these local disorders is given in Table 4.

A summary of these categories of airways infection–primary, opportunistic, and conditional–is given in Table 5.

NUMBERS OF MICRO-ORGANISMS

The number of micro-organisms involved is also important. A small load of primary pathogenic micro-organisms, as often occurs in influenza in individuals with rather normal immunological and local defense mechanisms, may result in subclinical disease, but a large load will almost certainly lead to symptoms. A similar phenomenon is recognized with respect to secondary pathogenic micro-organisms: a small load (e.g. of H. influenzae in the normal nasopharynx) usually does not result in bacterial infection of the lower airways during influenza, but a large load (sinusitis) may do so. Furthermore, immunologically compromised patients and patients with airways compromised by local disease are liable to become infected even when a rather low number of bacteria are present in the air they breathe. This means that individuals with impaired general or local defense mechanisms are more likely to develop bacterial complications during influenza virus infections than are individuals with an undisturbed defense system.

INFLUENZA VIRUS INFECTION; EARLY AND LATE BACTERIOLOGICAL COMPLICATIONS

Epithelial Destruction

Influenza virus invading and proliferating in ciliated epithelia cells is able to destroy a large part of the epithelial lining of the airways, thus providing an opportunity for common micro-organisms to penetrate and establish a bacterial infection, either bronchitis or pneumonia. This may occur within hours after the clinical onset of the influenza and is an early bacteriological complication. In this situation staphylococci are especially prevalent. Sometimes it may take several days for bacteria to colonize and penetrate, which leads to late bacteriological complications, in which H. influenzae and Strep. pneumoniae are often involved.

Studies during the pandemic of Asian influenza in 1957–1958 showed that it may take up to three weeks for the epithelium to recover, a period in which the airways are vulnerable to invasion by bacteria. From Fig. 1, which shows data on 148 influenza deaths in The Netherlands in 1957[2], it can be seen that 24% of the

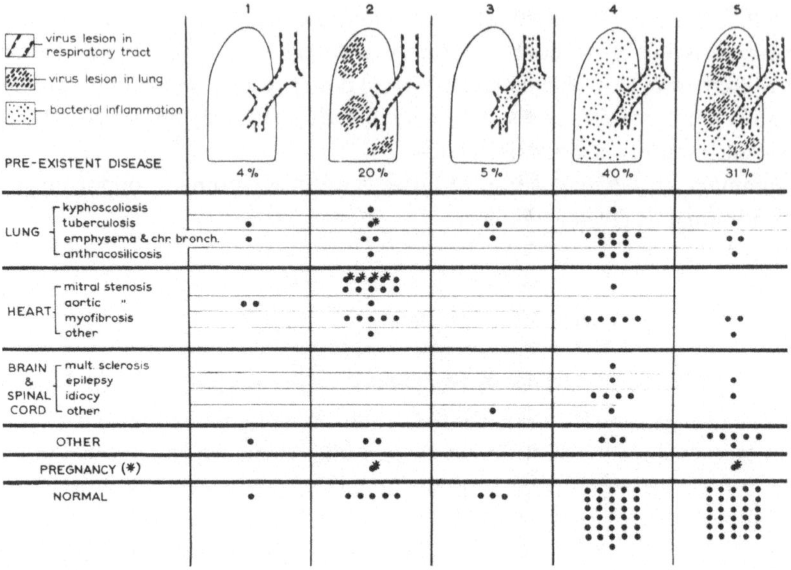

Fig. 1. Types of histopathological lesions in virologically confirmed cases (n = 148) of Asian influenza (July 1957-March 1958). From Hers et al., 1958 [4].

lethal cases concerned viral lesions only and the remainder bacterial infections as well. In patients with viral lesions pre-existent heart disease was prevalent and in those with bacterial complications diseases of the lung, heart, and central nervous system played an important role, but most of the individuals had previously been healthy.

Which Bacteria are Involved?
One might expect that only commensal floral normally present in the upper

Table 6. Bacterial flora found in sputum of 143 patients suffering from Asiatic influenza*.

Disease	Bacteria present					
	H. influenzae	Strep. pneumoniae	Strep. pneumoniae and h. influenzae	S. aureus	Strep. pyogenes	Others
Bonchitis	17	14	3	11		
Bronchopneumonia		8		44	2	4
Pneumonia		28	8			4
Total	17	50	11	55	2	8

* From Mulder and Hers, 1972 [2] and Goslings, 1958 [3].

airways would have an opportunity to colonize the damaged lower airways, having reached this area via the airstream or by epithelial spreading. The findings during the Asian pandemic indicate that this indeed happens (Table 6) [3]. In this respect an influenza virus infection closely resembles the situation seen in chronic obstructive pulmonary disease. In the latter, too, a local defect–very different in nature, but nevertheless a local defect–predisposes to infection by bacteria belonging to the common colonizers of the nose and throat, i.e., H. influenzae and Strep. pneumoniae. These bacteria tend not to be very aggressive and seldom lead to the early type of bacteriological complications unless the bacterial load is large. When, however, the patient happens to be a carrier of staphylococci, as are most patients with furunculosis and the relatives living with them, the situation is different. Staphylococci seem to possess more pathogenetic properties than the normal commensals in the upper airways do, and often play a nasty role in early bacteriological complications of influenza, leading to overwhelming broncho-pneumonitis, sepsis, toxic shock, and sometimes a lethal outcome within hours.

Treatment with Antibiotics

It seems to be generally accepted that patients with obvious bacterial complication during influenza should be given antibiotics. During a real influenza epidemic, a situation which could certainly occur in the coming years, any patient with sudden high fever, headache, sore throat, and a cough should be considered as having influenza, unless and until another cause of the condition becomes apparent. A bacterial infection is usually manifested by a productive cough and purulent sputum, at least in cases of infection by H. influenzae and pneumococci. A gram stain will indicate that bacteria are present and antibiotic treatment will be instituted. Sometimes, however, pure influenza virus infection without bacterial growth will produce purulent sputum due to high numbers of dead epithelial cells or fragments of such cells. In these cases it is difficult to discriminate between a non-bacterial and a bacterial infection, particularly because negative results of a bacteriological investigation do not rule out bacterial infection absolutely. In our experience, up to 25% of such routine investigations may prove to be false negative. I therefore think that we should treat such a patient with antibiotics if he is severely distressed.

In cases with certain or extremely likely staphylococcal involvement, another line has to be followed. In such cases we formerly administered a combination of high-dosage penicillin and an aminoglycoside by the parenteral route if we had an indication that the suspected staphylococcus was home-born, but cloxacillin and an aminaoglycoside when the staphylococcal strain might be hospital-born and thus probably resisant to penicillin. Because of the recently reported high percentage of penicillin-resistant staphylococcal strains among the home-born strains, however, we now use cloxacillin in all cases where staphylococcal involvement is suspected.

202

Prophylactic Administration of Antibiotics

A totally different situation is involved in the problem of advice on the prevention of bacterial complications of an influenza virus infection in patients already at risk of acquiring bacterial respiratory infections because of other reasons (Table 7). This point concerns the following groups of patients:

- Those with chronic presence of potential secondarily invasive microorganisms in the upper or lower airways, i.e., those suffering from chronic sinusitis, bronchiectasis, cystic fibrosis, and Kartagener's syndrome. In these patients the bacterial load is already heavy and epithelium damaged by viral infection will be highly susceptible to infection by the bacteria present.
- Patients in whom chronic obstructive airway disease, such as chronic bronchitis, persistent bronchial asthma, and emphysema, hampers drainage of airways, which may make them liable to bacterial infection. Influenza virus infection in this group may involve an extra risk.
- Patients with a deformed thoracic cage (scoliosis) or neurological abnormalities (poliomyelitis and other central or peripheral nervous disease) represent essentially the same situation as the foregoing group, having a deficient clearing mechanism because they cannot breathe properly.
- Patients with a cardiac disease, whether or not associated with pulmonary congestion, diabetes, or renal insufficiency, since these factors facilitate the development of viral lesions and a bacterial infection.
- Staphylococcus carriers, i.e., patients with furunculosis and relatives living close to them.
- Compromised hosts (immune deficiency).

These are the patients for whom vaccination against influenza is advised

Table 7. Indications for antimicrobial prophylaxis in influenza virus infection.

Normal individuals	No
Carriers of S. aureus	Yes
Persistent respiratory infections	Yes
Respiratory disease with recurrent and severe infections	Yes
Respiratory disease without recurrent and severe infections	No
Skeletal and neurological disease	No*
Cardiac disease	
Diabetes	No*
Renal insufficiency	No*
Immunodeficiency	No*
Pregnancy	No*

* Except when bacterial airway infections are known to have seriously disabled such a patient in the past.

officially. If, however, such patients develop clinical influenza, should they be given antibiotics even in the absence of signs of bacterial infection? This is a highly relevant question in practice, and we feel the risk associated with withholding antibiotics must be weighed against the (often more remote) disadvantages connected with a less discriminating use of antibiotics.

I think this decision should be taken rather individually. We would probably treat a patient known to have recurrent severe bronchial infections with an antibiotic, guided by the knowledge of the presence of bacteria in the past, but would not give antibiotics to an asthmatic who only shows signs of obstructive disease in the hay-pollen season, if he developed uncomplicated influenza in February.

Table 7 presents guidelines for the decision as to whether to prescribe antibiotics in patients with influenza who are at risk but do not yet show a bacterial infection.

Choice of Antibiotics

Table 8 gives a list of some of the above-mentioned conditions in which there is an indication for treatment with antibiotics, as well as the drugs which most Dutch physicians would prescribe at present.

Table 8. Antibiotics applied in influenza virus infection.

Micro-organism	Patient	Risk of infection	Infection present
S. aureus	carrier	cloxacillin (parenteral) (in case of allergy: erythromycin)	cloxacillin (parenteral)
Strep. pneumoniae	chronic respiratory	amoxycillin (oral) ampicillin (parenteral)	amoxycillin (oral) ampicillin (parenteral)
H. influenzae	disease	(in case of allergy: erythromycin, tetracyclin)	

REFERENCES

1. Orie NGM, Löwenberg A: Erreger, Wirt und Terrain: Über den Begriff der Pathologität in Bereich des Respiration tractus. Tatsachen und Hypothesen. Prax Pneumol 31:533, 1977.
2. Mulder J, Hers JFPh (eds.): Influenza. Wolters-Noordhoff Publ. Co., Groningen, 1972.
3. Goslings WRO Bacterial complications of influenza, prophylaxis and treatment. In: Influenza. Proceedings of a Boerhaave course for postgraduate medical education. Leiden, 1958, p. 48.
4. Hers JFPh, Masurel N, Mulder J: Bacteriology and histopathology of the respiratory tract and lungs in fatal Asian influenza. Lancet ii: 1141, 1958

DISCUSSION

Dr Mouton: I have two questions. The first concerns the remark you made about the 25% falsely negative cultured gram stains–I did not understand that properly–in cases of infections in the lungs. I do not think it concerns specifically staphylococcal infections in influenza, but whatever infection may be present. I quite agree with you–and we discussed it quite often–that the bacteriology of sputum is not very reliable and a lot of mistakes are made in laboratories, especially by regarding non-pathogens as pathogens in a lot of cases. But the figure of 25% for falsely-negative is a different matter. I cannot believe that. I want to know whether you have documented data to show this.

The second question regards the matter of Staphylococcal aureus carriers. Everybody would agree that anybody who is prone to staphylococcal infection would need prophylaxis in a situation of influenza. But, when you mention staphylococcal carriers, I think it is rather difficult, because 40% of us are staphylococal carriers. Do you not mean those people who have recurrent staphylococcal infections, particularly skin infections?

Dr Dijkman: With regard to the first question: the data which led me to give the figure of 25% comes from the bronchitic out-patient department in our hospital, where we had clinical evidence to accept a bacterial infection in exacerbations of chronic bronchitis. In about 45 of such exacerbations, bacteriological confirmation was obtained in about 70–75%. That means that 35% of the patients whom we considered having an infection because of purulent sputum and/or reaction on antibiotic treatment, were given a false negative report. In this group we had no Staphylococcus aureus infections. Because all had chronic bronchitis and were producing sputum continually, it was sometimes difficult to judge whether the sputum was mucous or purulent.

Dr Mouton: I would like to continue on this subject. I have the feeling that your remark may have been misunderstood. I quite agree that, during therapy, in the course of a bronchitis you sometimes have negative sputa at times when the sputum may even be purulent, or slightly purulent. Is it not so that these infections have actually abated or that anti-bacterial therapy makes culture of the bacteria impossible?

Dr Dijkman: In general that could well be. I agree with what you say. It is possible that more factors play a role, but in our group of bronchitis sputum culture was performed before antibiotics were given. You stated that 40% of the population could be considered staphylococcal carriers. If you make cultures of several parts of the body, perhaps. But I would consider a staphylococcal carrier as an individual of whom I know–unless it is a doctor or surgeon who is repeatedly nose-swabbed–that he is suffering from staphylococcal skin lesions or has relatives with these lesions living very close to him. These are what I would consider staphylococcal carriers; that is not 40% of the population.

Dr Mouton: I think we are of the same opinion, but your definition is not quite the same as to what is meant to be a staphylococcal carrier.

Prof. Grob: Perhaps one could sketch what an epidemic of influenza really is like in a community. The last epidemic was in 1968 when the new Hongkong strain came. At that time, I was doing something like 80 home visits a day, which is about 10 or 12 hours' work, at which time of course it is really impossible to organize any sort of bacteriology, you will appreciate. Two waves of death came with that epidemic. One, fortunately in the old and the infirm, where some died within 12 hours of contracting the disease, sometimes before I got around to seeing them. There was a second wave of death, about a fortnight to three weeks later. Here, the postmortems showed the changes of large Staphylococcal pneumonia. The policy we then adopted–this is a point in question whether it was a wise one–was to give everybody who contacted us and who seemed to be vulnerable, tetracyclines. It was all we could manage.

Would you perhaps change that policy of 12 years ago, now? What should we be thinking about for the next epidemic?

Dr Dijkman: It is surprising to me, but my experience goes exactly the other way round: people getting into trouble shortly after contracting influenza, in my view, tend to be infected with staphylococci and, later on with pneumococci or Haemophilus influenzae.

As to what you did in dealing with staphylococci: give tetracycline to relatives, whether they have influenza or not, as a prophylactic measure. Do you really advocate this?

Prof. Grob: We gave it to anyone who called in. That is all you can do when your phone is ringing constantly. The local pathologists did 168 postmortems that week, so it may be that the bacteriology was not quite of the high, excellent quality that one would hope. It is very difficult to know just what practical work you can do when you are under these 'siege' conditions.

Dr Dijkman: I recognize the difficult situation. I had the same experience in 1958–I had just graduated–when the second wave of Asian influenza was around and the general practicians had a hard time. For the moment I would think that confronted with the risk of staphylococcal complications, I would prescribe cloxacillin to a lot of people.

Chairman: I think if you give only cloxacillin the first day, you will be giving enough for the pneumococci.

Dr Thompson: As a choice of treatment for staphylococcus infections, in cases of penicillin allergy, you advocate erythromycin. I wonder why not one of the cephalosporins?

Dr Dijkman: I think it is a matter of taste, I always have felt that giving cephalosporins would bring a somewhat larger risk on side-effects. Not because the anti-staphylococcal activity will be so much higher, but the balance between the complications of the drug therapy with erythromycin compared to cephalosporins–in my limited experience at least–would favor erythromycin.

Dr Van Boven: I would like to elaborate on the sputum culture, 20 to 25% is false negative. We all know the problem of contamination. How much should that be? Perhaps 50% of the cultures we make do not give the information that you might expect. You cannot tell from the cultures results if these results are true or false. So, what are we really doing? I should like to ask clinicians. What do you want to do with these laboratory results?

What is the relevance of microscopy of the sputum? Dr Dijkman alluded to that in saying we should look at the sputum macroscopically and under the microscope. Is that not the solution of the whole problem? Or does it provide too many uncertainties?

Dr Dijkman: I said you had to be careful about the culture of the sputum: beware of a false negative result. What we expect from a culture, as clinicians, is confirmation of the Gram stain, confirmation of a clinical impression, and the possibility to have a sensitivity test of the isolated organism.

Dr Van Boven: What we heard yesterday is to use cloxacillin against Staphylococcus aureus infections. There is a very low frequency of meticillin resistance which you should have to take into account. Haemophilus influenzae and Streptococcus pneumoniae are still sensitive to the drugs you use. Although in the monitoring aspect, I can agree that you should take samples to determine resistance, but in your day-to day practice this has no value.

Dr Mouton: I think Dr Van Boven has a very pessimistic view of the whole thing, although I must agree that it is one of the subjects which is difficult to work on, because you make a lot of mistakes. With regard to the number of falsely negative or falsely positive cultures, I think they are somewhat exaggerated. I feel that falsely negative is rare and falsely positive indeed occurs in about 30% of the cases. I find that Dr Dijkman's reasoning is not very consistent with regard to the sensitivity testing, because when you know that 30% is not reliable in the culture, then the sensitivity test is not reliable for therapy.

Dr Van Boven: I would agree with the consequences of this: you should choose your therapy in the normal cases of bronchial infections and pneumonia on the basis of the clinical symptoms and the expected bacteria, and not on the basis of the small percentage of resistant strains reported in the literature, which is not relevant to therapy.

Dr Stam: Is there a place in Dr Dijkmans scheme for prophylactic treatment, very intensive, very short?

Dr Dijkman: What do you call prophylactic treatment? You could call treatment prophylactic when antibiotics are administered in a situation in which somebody has contracted clinical influenza while not yet showing signs of a bacterial infection. There are such situations. For instance, patients with kyphoscoliosis in which you know from experience that the patient will be in serious trouble in case of an infection. In this patient, you would try to prevent a possible bacterial complication. This is always an uncertainty: when should you do it and when not? This has very much to do with the individual patient. Patients who are vulnerable, you are more likely to treat. With patients who have objections against all sorts of antibiotics, you would probably be more restrictive.

Dr Stam: But, that is not my question. Do you treat these patients 7 days with 2 grams of ampicillin or do you give it for 2 days?

Dr Dijkman: If you do something, you had better do it well. I would give it at least a week.

17. ANTIMICROBIAL TREATMENT OF LEGIONELLA PNEUMONIA

PIETER L. MEENHORST

INTRODUCTION

Legionnaires' disease entered the medical universe when a mysterious outbreak of pneumonia affected 149 persons attending the 58th convention of the American Legion's Pennsylvania Department in Philadelphia in July, 1976. The respiratory illness, which was soon called legionnaires' disease by press and public alike, had a broad range of manifestations from a mild respiratory sickness to a multisystem affection involving extensive pneumonia, gastro-intestinal symptoms, confusion, hepatic dysfunction, shock, and renal failure.

The illness was not confined to convention participants. An additional 72 cases were discovered. All together, 221 persons had become ill; 34 of them died and of the 123 hospitalized patients, 111 had radiographically proven pneumonia [1,2].

In January, 1977, the Center for Disease Control (Atlanta, USA) reported the isolation of a Gram-negative bacterium from embryonated yolk sacs inoculated with spleen homogenates of guinea pigs infected with lung tissue of disease victims of the Philadelphia epidemic [3]. In fact, the agent of what then was called the legionnaires' disease was discovered by a procedure designed to isolate rickettsia. Comparison of morphological and biochemical data by Brenner et al showed no relation to known bacterial species. They proposed the name Legionella pneumophila for this previously unrecognized species [4]. Using L. pneumophila as an antigen, McDade developed an indirect immunofluorescent antibody test [1]. The serum of most of the patients who survived the Philadelphia epidemic showed a significant rise of antibody against L. pneumophila.

Epidemics Diagnosed Retrospectively
Earlier mysterious outbreaks of pneumonia could be explained by serological testing of stored sera for L. pneumophila. An outbreak of pneumonia in a large psychiatric hospital in Washington, DC, in July, 1965, had many clinical features in common with the Philadelphia epidemic, and was shown in 1977 to have been caused by L. pneumophila or an antigenically similar micro-organism [5].

The epidemic in Pontiac, Michigan, in July of 1968, where at least 144 persons visiting or working in a county health department building developed an acute febrile affection, later called 'Pontica Fever', had a different clinical picture [6]. The major symptoms being myalgia, fever, chills, and headache. Respiratory symptoms were absent or minimal. In 32 cases serological evidence of infection with L. pneumophila was obtained, and this bacterium was isolated from the lungs of guinea pigs exposed in the building in Pontiac at the time of the outbreak as well as to aerosols of water from the air-conditioning system [7]. This epidemic differed from the one in Philadelphia as to incubation period, attack rate, and clinical illness. The incubation period was 24–48 hr versus 2–10 days, the attack rate was 95% versus an estimated 1.6% in the Philadelphia epidemic, and, above all, the sickness was self-limiting, lasting from 2–7 days, and was not associated with pneumonia or death.

The following conclusions can be drawn from these three epidemics. Not all of the patients fulfilling the criteria for legionnaires' disease in the Philadelphia epidemic had roentgenologically proven pneumonia. The clinical illness caused by L. pneumophila does not necessarily involve pneumonia and can be self-limiting. L. pneumophila was only isolated from patients with pneumonia. And, an airconditioning system may have been responsible for 'Pontiac fever' by distributing aerosols contaminated with L. pneumophila.

Sporadic Cases Diagnosed Retrospectively
A number of sporadic cases have been diagnosed retrospectively, most of them by serological testing of stored sera [8,9,10]. Thus, the disease does not always lead to an epidemic.

EPIDEMIOLOGY

In the last three years much more has been learned about legionnaires' disease. Six serogroups of L. pneumophila have been isolated, not only from patient material but in some cases also from their immediate environment [11,12]. L. pneumophila has been isolated from soil, water from cooling equipment, and from many surface waters within the USA [13,14]. Thus, it seems to be an ubiquitous micro-organism. The mode of transmission of L. pneumophila in many outbreaks and in sporadic cases is unknown. The only documented mode of spread of L. pneumophila is airborn. In the USA more outbreaks as well as over a thousand sporadic cases have been diagnosed [15]. A relatively small outbreak has been reported from Great Britain [16]. An increasing number of sporadic cases have been reported from Australia, Austria, Belgium, Canada, Denmark, France, Germany, Israel, The Netherlands, Spain, Sweden, Switzerland, and Great Britain.

In different studies an association between legionnaires' disease and travel in the southern European countries has been found [10,17,18]. Travel in endemic areas and diminished host defenses due to heavy smoking, excessive alcohol ingestion and overexposure to sun during the holidays are possible, but not substantiated, explanations for this association [18]. Both sporadic cases and outbreaks occur throughout the year, but appear to be more common between June and November [15]. Seasonal factors seem to influence the occurrence of legionnaires' disease and may differ geographically. The incidence of legionnaires' disease is not known and will probably differ from country to country, and from area to area. The incidence of sporadic legionnaires' disease has been estimated as 12 cases/100 000 population/yr for the USA [19].

Recently, the largest outbreak of legionnaires' disease outside the USA occurred in Västerås, Sweden, where between August 28th and September 21st, 1979, 66 residents and one visitor developed symptoms like those of the epidemic in Philadelphia. Serological evidence indicating an infection with Legionella serogroup 1 was found [20]. This serogroup was isolated from patients and water condensate on the roof of an shopping centre, which had been visited by the large majority of ill persons.

Nosocomial Cases

Cases of nosocomial legionnaires' disease have been recognized in a number of centres [21,22]. Especially patients with decreased host resistance seem to be at risk.

It it likely that in some hospitals these nosocomial cases represent exogenous infections with L. pneumophila and not reactivation of previous infections. Since March, 1977, 180 cases of nosocomial legionnaires' disease have been diagnosed in the Wadsworth Veterans Administration Hospital in Los Angeles ([22]; Dr. R.D. Meyer, Wadsworth Veterans Administration Hospital, Los Angeles, USA, personal communication). Various serogroups of L. pneumophila have been isolated from patients and from cooling towers. Even after adequate decontamination of these cooling towers, cases of nosocomial legionnaires' disease continue to occur [11]. Nosocomial cases of legionnaires' disease have also been reported from Great Britain, The Netherlands, and Yugoslavia [10,12,23].

Thus, what initially appeared to be a rare disease may well turn out to be one of the major bacterial pneumonias, particularly in the immunocompromised patient. The role of L. pneumophila in respiratory-tract infections without pneumonia is not known, but it must be kept in mind that cases of legionnaires' disease without pneumonia have been identified. Seroconversion for L. pneumophila has been observed in asymptomatic patients admitted to the same hospital for different reasons [24]. Moreover, the clinical spectrum of the disease remains to be determined; legionnaires' disease is not a clinical entity. With respect to nomenclature, the disease caused by L. pneumophila, in which the pneumonia is a prominent feature, might more appropriately be called Legionella pneumonia, because the latter seems to be one of the manifestations of Legionella infections.

Recently, pneumonias caused by Gram-negative organisms similar to L. pneumophila have been described [25,26].

CLINICAL ASPECTS OF LEGIONELLA PNEUMONIA

Clinical Picture
The signs and symptoms of Legionella pneumonia are summarized in Table I.

Table 1. Common symptoms and signs in patients with Legionella pneumonia.

Prodromal state (1–4 days)	During development of lung infiltrate	During progression of the disease
Malaise	Cold chills	Mental confusion
Anorexia	High fever (> 38.5 °C)	No sputum or small amounts
Myalgia	Diarrhoea	
Headache	Abdominal pain	Chest pain
Diarrhoea	Nausea	
	Vomiting	
	Dry cough	
	Chest pain	

For sporadic cases, the incubation period is still unknown. In the Philadelphia epidemic the incubation time was 2–10 days. In nosocomial cases of Legionella pneumonia the duration of hospitalization before the onset of symptoms has varied from 6 days to 6 months [22]. In the series comprising the first 453 sporadic cases in the USA reported by the Center for Disease Control, males predominate over females by three to one [15]. Later and smaller studies showed a varying degree of male predominance.

The age range of the patients has been 10 months to 84 years, with a mean of 54 years for males and 56 for females [7]. Typically, Legionella pneumonia occurs in middle-aged or older individuals.

The majority of cases of Legionella pneumonia include a history of cigarette smoking. In many patients this is the only predisposing factor that can be demonstrated. In the typical case, initial symptoms of Legionella pneumonia include malaise, anorexia, myalgia, and headache, followed within 48 hr by a high fever (> 38.5°C), often preceded by recurrent rigors.

Typical Case of Legionella Pneumonia
The typical case is exemplified by the case history of the first patient in whom we

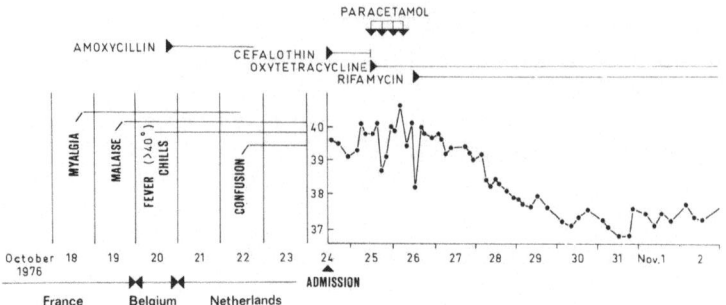

Fig. 1. Some details of a typical case of Legionella pneumonia. From PL Meenhors et al., 1978 [60].

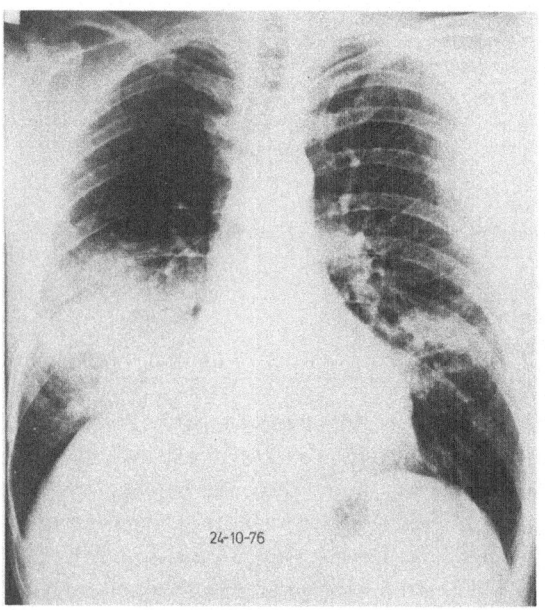

Fig. 2. Chest-X-ray: 24–10–1976. From PL Meenhorst et al., 1978 [60].

diagnosed Legionella pneumonia, a 49-year-old man who became ill on October 18th, 1976, after a holiday in France. Initially, he complained of malaise and myalgia, which was followed by cold chills and then fever (see Fig. 1). He was treated with amoxicillin by his physician. On October 24th the patient was referred to us because his condition had deteriorated and pneumonia was suspected.

On admission, the physical examination showed a very seriously sick, dyspnoeic, and mentally confused man. The respiration rate was 50/min, and central cyanosis was present. The rectal temperature was 40.6 °C the pulse rate 140/min.

Positive findings were otherwise confined to the chest. Dullness to percussion was present in the right lower lung field posteriorly, and coarse inspiratory rales were heard. A few days later, bronchial breath sounds were heard. Abnormal laboratory data: ESR: 87 mm. Leucocytes $5.6 \times 10^9/l$. Blood smear: 50 band forms. Blood gas analysis: pH 7.5, pO_2 54 mm Hg, PCO_2 23 mm Hg, HCO_3 18mmol/l. Sodium 123 mmol/l. Creatinine: 135 μmol/l. Chest X-ray: see Fig. 2.

Clinical Course. In the scarcely produced sputum a few leucocytes and Gram-positive cocci were found on the Gram smear. Initially, we considered as likely a streptococcal or staphylococcal pneumonia secondary to a viral upper respiratory tract infection. Pneumonia due to aspiration or Klebsiella pneumonia seemed less likely. When no clinical response to treatment with cefalothin was seen within 24 hr and no sputum was produced, we added mycoplasma and chlamydiae to the list of possible causes of the pneumonia. Oxytetracycline was given, and rifamycin was added when a few colonies of Staphylococcus aureus were cultured from a sputum sample. However, at this stage of the disease we did not believe that the pneumonia was caused by this micro-organism. On the fourth day after admission the temperature dropped and the clinical condition improved gradually. For more than a week high doses of oxygen were needed and the patient was mentally confused. Initially, pulmonary infiltration was progressive (see Fig. 3).

Radiographically, resolution of the lung infiltrate took more than six weeks (see Fig. 4). The patient recovered completely. We could not establish the cause of the pneumonia. Retrospectively, the diagnosis Legionella pneumonia was made on the basis of a significant rise of antibody titre against L. pneumophila serogroup 1.

Relative bradycardia has been reported [22]. In this stage one or more of the gastro-intestinal symptoms, which include abdominal pain, diarrhoea, nausea, and vomiting, may be present. However, diarrhoea may precede fever or systemic symptoms by one to six days [22].

Most patients develop a moderately dry cough on the second or third day after the onset. Small amounts of mucoid or minimally purulent sputum are produced by about half of the patients; this may change into mild or moderate haemoptysis during the course of the disease. Chest pain, often pleuritic in origin, may precede abnormalities found on physical examination of the chest. Inspiratory rales may be heard as the first sign of lung involvement. Later in the course, consolidation or a pleuritic friction rub is found on physical examination. In a few patients signs of pleural effusion may precede signs of parenchymal infiltrate. However, when the patient is seen in a relatively late stage of the disease, signs of pleural effusion may predominate. Central nervous system symptoms consisting of confusion, delirium, and dysarthria are frequently present. In these patients signs of meningeal irritation are lacking. Other neurological findings observed in

214

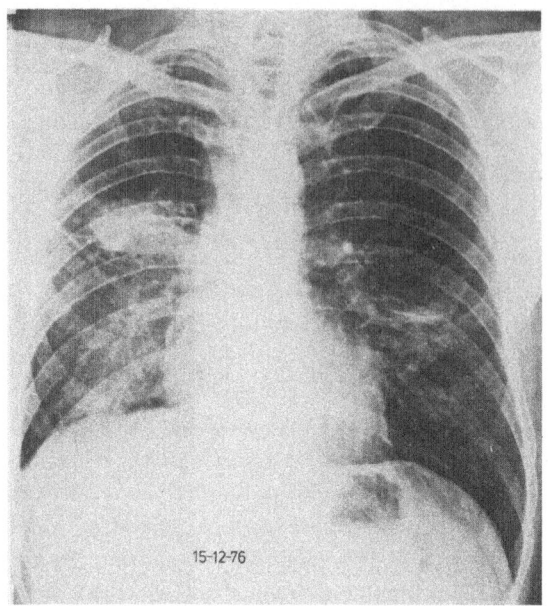

Fig. 3. Chest-X-ray: 15–12–1976.

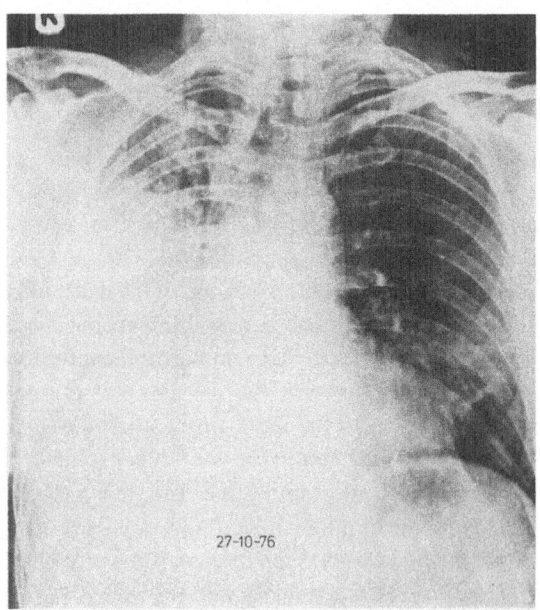

Fig. 4. Chest-X-ray: 27–10–1976.

a few patients include coma, cranial nerve palsy, aphasia, extensor plantar response, ataxia, disturbances of sensation, isolated grip and limb weakness, and seizures [27, 28]. A high proportion of the mentally confused patients show amnesia for the acute phase of the disease. Without appropriate antimicrobial treatment a high fever will persist. Spontaneous resolution of the infection may start on the 8th to 10th day of the illness, which resembles pneumococcal pneumonia in the pre-antibiotic era.

Laboratory Findings

The laboratory findings are summarized in Table 2. The sedimentation rate is elevated. Leucocytosis, accompanied by neutrophilia and lymphopenia, occurs in various degrees [29]. Leucopenia has been found in a minority of the cases and was associated with a poor prognosis [1]. Isolated transient thrombocytopenia and thrombocytopenia due to diffuse intravascular coagulation have been seen in a few patients [10, 27, 30]. Pancytopenia with a hypoplastic bone marrow has been reported, but there is no convincing evidence that this was due to the infection with L. pneumophila [31].

Mild proteinuria is present in most patients, whereas microscopic haematuria has been found in about 10% of the patients. Heamaturia is significantly more common in patients who develop renal failure than in those whose renal function remains normal [32]. Elevation of serum creatinine and blood urea nitrogen has been reported frequently, and cases of renal failure have been described [1, 32, 33, 34]. Elevation of the levels of alkaline phosphatase, serum glutamioxaloacetic

Table 2. Abnormal laboratory findings in patients with Legionella pneumonia.

Elevated erythrocyte sedimentation rate (often > 70 mm)	+
Granulocytosis (varying degree of shift to the left)	+
Granulocytopenia	−
Thrombocytopenia	−
Diffuse intravascular coagulation	−
Mild proteinuria	+
Microscopic haematuria	−
Elevated serum creatinine	+
Renal failure	−
Hyponatraemia	+
Hypophosphataemia	+
Hypoalbuminaemia	+
Elevated alkaline phosphatase	+
Hyperbilirubinaemia	+
Elevated SGOT	+
Elevated SGPT	+
Elevated CPK	−

+: common finding.
−: in less than 50% of the cases.

transaminase (SGOT), serum glutamic pyruvic transaminase (SGPT), and bilirubin, reflecting hypatic dysfunction, is present in most cases to a varying degree. In one series the incidence of a low serum albumin level (≤ 25 g/l) was remarkably high on admission [29]. Hyponatraemia (< 130 mmol/l), which is commonly present, has been ascribed to the syndrome of inappropriate antidiuretic hormone secretion [35]. Hypophosphataemia (< 0.81 mmol/l), typically occurring during the first 72 hr has been found in as many as half of the patients [35]. Rises in creatinine phosphokinase have been associated with rhabdomyolysis [36]. No abnormalities of the protein and glucose contents of the cerebrospinal fluid have been found, but a mild pleiocytosis (mainly monocytes and lymphocytes) may be present [10, 32].

Radiologic Findings

Unilateral involvement of the lung is found in the majority of the cases. Poorly marginated, round opacities located either centrally or peripherally and diffuse patchy bronchopneumonia are early radiologic findings [37].

As the disease progress, infiltrates increase and may extend over a whole lobe, showing a ground-glass appearance of dense alveolar consolidation, even after so-called effective antibiotic therapy has been started. Multilobar involvement is common. Pleural effusions are seen in as many as 50% of the patients [37]. In rare cases the pneumonia may progress to cavitation [10, 38, 39, 40].

Diagnosis

In sporadic cases it may be difficult to establish the clinical diagnosis Legionella pneumonia. Initially, pneumonia is often not suspected from the results of the physical examination. However, the absence of sputum or the small amount produced by a patient with pneumonia showing the above-described clinical and laboratory features may lead the physician to consider the diagnosis Legionella pneumonia. The differential diagnosis must include not only other bacterial types of pneumonia but also Mycoplasma pneumoniae, psittacosis, Q-fever, and viral pneumonias [41]. Additional support for the diagnosis Legionella pneumonia is provided by the absence of bacteria on a Gram smear as well as negative routine cultures of expectorated sputum, transtrachial aspirate, or bronchial washings, pleural fluid, lung aspirates, and lung biopts. Rapid diagnosis of Legionella pneumonia can be made by DFA examination of these materials [42, 43]. The Gram stain is not very useful. Rarely, faintly staining Gram-negative bacilli are seen. For fresh biopsy material or bronchoscopic washings the Gimenez stain is useful, but is not specific for L. pneumophila. When present in clinical specimens, L. pneumophila can be isolated on Feeley-Gorman agar and charcoal yeast extract agar supplemented with ferric pyrophosphate [44, 45]. L. pneumophila has also been isolated from blood [46, 47]. Most cases of Legionella pneumonia have been confirmed retrospectively by a significant rise in titre in the IFA test applied to serum obtained at the onset of the disease and to reconvalescent serum [41].

SUSCEPTIBILITY OF L. PNEUMOPHILA TO ANTIMICROBIAL AGENTS

For the choice of antimicrobial therapy, recommendations were originally based on retrospectively collected data on the Philadelphia outbreak of legionnaires disease, which showed that the case: fatality ratio was lowest in the patients treated with erythromycin or oxytetracycline [1]. Patients with Legionella pneumonia treated with penicillins or cephalosporins, alone or combined with aminoglycosides, showed no clinical improvement. However, Thornsberry's in vitro studies indicated that nine strains of L. pneumophila are susceptible to many antimicrobial agents [48] (see Table 3). The MBC's were generally the same or one dilution higher than the MIC's. All strains produced a β-lactamase that was basically a cephalosporinase but also had some activity against all the penicillins tested [48]. These in vitro data differ from the results obtained in a guinea pig model by Fraser et al. Erythromycin and rifampin prevented death of guinea pigs infected intraperitoneally with a lethal dose of L. pneumophila isolated from a victim of the Philadelphia epidemic, whereas penicillin, chloramphenicol, tetracycline, and gentamicin gave no significant effect [49].

Table 3. Minimal inhibitory concentrations (MIC) and minimal bactericidal concentrations (MBC) of 15 antimicrobics for nine strains of Legionella pneumophila; obtained in FG cysteine iron broth.°

Antimicrobic	MIC		MBC	
	Mode (μg/ml)	GM* (μg/ml)	Mode (μg/ml)	GM (μg/ml)
Penicillin	2.0	1.0	2.0	1.0
Ampicillin	0.25	0.5	0.25, 1.0	0.5
Cephalothin	8.0	8.0	8.0	8.0
Cefoxitin	0.12	0.12	0.12	0.12
Tetracycline	8.0	4.0	4.0	4.0
Minocycline	0.25, 0.5	0.5	1.0	0.5
Doxycycline	2.0	2.0	2.0	2.0
Gentamicin	0.06	0.06	0.12	0.12
Tobramycin	0.12	0.12	0.25	0.25
Amikacin	0.25	0.25	0.25, 0.5	0.5
Chloramphenicol	0.5	0.5	0.5	0.5
Erythromycin	0.12	0.12	0.12	0.25
Rifampin	$\leqslant$0.01	$\leqslant$0.01	$\leqslant$0.01	$\leqslant$0.01
Colistin	0.5	1.0	0.5, 1.0	1.0
Cotrimoxazole	1.2/0.06**	1.2/0.06	38/2.0	38/2.0

* GM = geometric mean.
** Sulfamethoxazole/trimethoprim (1 + 19 combination).
° Modified from Thornsberry et al., 1978 [48].

Lewis et al. determined the susceptibility of L. pneumophila to ten antimicrobial agents by inoculating embryonated eggs via the yolk sac [50]. When these drugs were administered prophylactically, the minimal dose preventing death was 0.02 mg for rifampin, 0.25 mg for gentamicin, 0.39 mg for streptomycin, 0.62 mg for erythromycin, 1.56 for sulfadiazine, 2.50 mg for chloramphenicol, and 20.0 mg for cephalothin. Smaller amounts delayed death, and lager or equal amounts rendered the embryos free of infection. In a dose of 5.0 mg. oxytetracycline protected 80% of the embryos from death, whereas 0.31 mg delayed death. Chlortetracycline in a dose of 0.50 mg was ineffective, as was 10.0 mg ampicillin. Administration of rifampin and erythromycin 72 hr after infection, at double the minimal prophylactic dose preventing death, resulted in survival of all embryos, and the same holds for gentamicin, streptomycin, sulfadiazine, and chloramphenicol when given 48 hr after infection (see Table 4). No bacteria could be demonstrated in embryos treated with rifampin or gentamicin 48 hr after infection or with erythromycin as late as 72 hr after infection.

L. pneumophila is frequently found within macrophages in lung specimens from fatal cases of Legionella pneumonia and within peritoneal macrophages in experimentally infected guinea pigs [51, 52]. This suggests that the microorganism behaves like a facultative intracellular micro-organism. Experiments carried out by Horwitz and Silverstein showed that L. pneumophila multiplies intracellularly and that it escapes the killing mechanism of human monocytes [53]. It is conceivable that the differences in the efficacy of the various antimicrobial agents are partially due to differences in their capacity to penetrate phagocytic cells.

Furthermore, it has been shown that L. pneumophila can lose virulence by

Table 4. Comparative efficacy of antimicrobial agents for delayed treatment of infection with Legionella pneumophila I in embryonated eggs.°

Antimicrobial agent	Maximum delay possible before treatment of infected embryos to attain:			
	Dose*	100% survival	Bacterial clearance	Increased survival time**
	(mg/egg)	(hr)	(hr)	(hr)
Rifampin	0.04	72	48	96
Gentamicin	0.50	48	48	72
Streptomycin	0.78	48	<4	72
Erythromycin	1.24	72	72	96
Sulfadiazine	3.12	48	4	72
Chloramphenicol	5	48	4	72

* Twice the amount of drug preventing all deaths when used prophylactically.
** Survival time at least 2 days longer than that of untreated embryos.
° Modified from Lewis et al., 1978 [50].

multiple passage on artificial media [54]. Although this phenomena has not been described for the more recently developed artificial media, caution must of course be exercised in interpreting results of in vitro susceptibility testing for L. pneumophila when data on the virulence of the strains in question are not available.

ANTIBIOTIC THERAPY FOR LEGIONELLA PNEUMONIA

Erythromycin has been advocated as the drug of choice for suspected Legionella pneumonia, and erythromycin combined with rifampin for confirmed cases not responding to erythromycin alone [41]. The efficacy of erythromycin for the treatment of Legionella pneumonia has been suggested by the following studies.

In Vermont, Va. (USA) 56 cases of Legionella pneumonia were diagnosed during a period of 5 months [55]. The disease was fatal in 3/6 (50%) untreated patients, in 6/24 (25%) patients given cephalosporin in 7/22 (32%) given gentamicin, in 6/23 (26%) given penicillin, and in 1/16 (6%) given erythromycin. Of the patients treated with erythromycin, 7 (44%) were immunocompromised, whereas 11 (46%), 8 (36%), and 6 (26%) of those given cephalosporin, gentamicin, and penicillin, respectively, were immunocompromised. The outcome in the erythromycin-treated patients was not significantly better.

Considerable experience in treating Legionella pneumonia was accumulated by Kirby et al. In their hospital, 65 cases of legionnaires' disease were diagnosed between May, 1977, and December 15, 1978 [22]. The over-all case: fatality ratio was 25% (16/64), for the erythromycin-treated group the ratio was 13% (6/46), and for those not given erythromycin 55% (10/18). In the group of immunocompromised patients treated with erythromycin the case: fatality ratio was 24% (4/17) versus 80% (8/10) when erythromycin was not included in the therapy [22]. Of the 42 patients who received erythromycin for more than 24 hr and whose response could be evaluated, equal numbers initially received the drug orally and intravenously [22]. Five patients initially received erythromycin 4 g/d, the others 2 g/d. No difference in response was noted between the two dosage regimens, and the differences in response between the oral and the intravenous groups were small. However, no criteria are mentioned for the patients condition of the illness at the start of erythromycin therapy. It may well be that a bias was introduced via the selection of patients for the different doses and routes of administration.

In the critically ill patient these antibiotics should be administered parenterally. It has been shown for various antibiotics that absorption may be impaired in disease, resulting in serum levels lower than desired. In Legionella pneumonia one cannot take the risk of therapy failure due to impaired absorption of the drug. Moreover, it is possible that high levels of erythromycin are bactericidal for L. pneumophila, it has been shown that erythromycin in high concentrations is bactericidal in vitro and it might also be bactericidal in vivo [56].

Recently, we obtained excellent results in five severely granulocytopenic patients (granulocytes < 500/mm³) with Legionella pneumonia. When these patients developed pneumonia no sputum was produced, and the bacteriological data were not indicative for the cause of the pneumonia. A presumptive diagnosis Gram-negative pneumonia was made in four of these patients. Legionella pneumonia was included in the differential diagnosis. Therapy consisted of erythromycin 4 g/d i.v. and tobramycin i.v. in three cases and erythromycin 4/g i.v. and cefamandol i.v. in one patient. One patient was treated with erythromycin 4 g/d i.v. alone, because L. pneumophila seemed to be the most probably micro-organism responsible for the clinical picture. Tobramycin and cefamandol are thought to be ineffective for the treatment of Legionella pneumonia. Although we cannot exclude a beneficial role of these drugs in the recovery of our patients, we attribute the clinical response mainly to erythromycin. In the absence of granulocytes it is difficult if not impossible to overcome a serious infection with bacteriostatic drugs alone. In view of the dramatic response on erythromycin therapy in these granulocytopenic patients, we speculate that erythromycin given in high doses i.v. may be bactericidal for L. pneumophila.

No well-documented data are available on the influence of the duration of erythromycin therapy. Relapse or protracted convalescence has been noted in a few patients treated for less than 3 weeks [38].

Recovery from Legionella pneumonia has been related to therapy with many antimicrobial agents. In the Philadelphia outbreak of legionnaires' disease, patients treated with oxytetracycline and erythromycin had the lowest case: fatality ratio [1]. Some patients have been described who showed improvement on therapy with doxycycline [57, 58]. It may well be that the more lipid-soluble tetracyclines such as doxycycline and minocycline are effective when given early in the course of Legionella pneumonia. In general practice, erythromycin and doxycycline are widely used, especially for the treatment of respiratory-tract infections. It is conceivable that due to the use of these antibiotics, many patients will not develop a serious Legionella pneumonia. On these grounds it seems probable that the role of L. pneumophila in respiratory-tract infections is underestimated at present.

Rifampin seems to be a very potent drug for the treatment of Legionella pneumonia. When rifampin is given alone for the treatment of infections, one-step resistance may occur. Since all isolated strains of L. pneumophila are susceptible to rifampin, one should not take the risk of inducing resistant strains by using rifampin alone. In nontuberculous conditions a short course of rifampin in combination with another antimicrobial drug does not contribute to the emergence of rifampin-resistant strains of Mycobacterium tuberculosis or Gram-negative bacteria [59]. Thus, in a case of Legionella pneumonia not responding to erythromycine alone the addition of rifampin to the therapeutic regimes deserves consideration.

After ten days of treatment with erythromycin L. pneumophila has been isolated from a patient with Legionella pneumonia [43]. In two immunocompromised patients with Legionella pneumonia L. pneumophila was demonstrated in the lung by DFA after two weeks of treatment with erythromycin i.v. [38]. No cultures were made in these two cases. According to the available data, erythromycin seems to be the drug of choice for the treatment of Legionella pneumonia. In immunocompromised patients, especially those treated with corticosteroids, the addition of rifampin to the therapy early in the disease should be considered.

Controlled prospective clinical trials are needed to evaluate the efficacy of various antimicrobial agents not only in terms of survival and clinical response but also as to long-term sequelae.

REFERENCES

1. Fraser DW, Tsai TF, Orenstein W, Parkin WE, Beecham HJ, Sharrar RG, Harris J. Millison GF, Martin SM, McDade JE, Shepard CC, Brachman PS, the field investigation team: Legionnaires' disease: Description of an epidemic. N Eng J Med 297:1189, 1977.
2. Fraser DW, McDade JE: Legionellosis. Sc. Am. 241:82, 1979.
3. McDade JE, Shepard CC, Fraser DW, Tsai TF, Redus MA, Dowdle WR, the laboratory investigation team: Legionnaires' disease: Isolation of a bacterium and demonstration of its role in other respiratory disease. N Eng J Med 297:1197, 1977.
4. Brenner DJ, Steigerwalt AG, McDade JE: Classification of the Legionnaires' disease bacterium: Legionella pneumophila, genus novum, species nova, of the family Legionellaceae, familia nova. Ann Intern Med 90:656, 1979.
5. Thacker SB, Bennet JV, Tsai TF, Fraser DW, McDade JE, Shepard CC, Williams KH Jr, Stuart WH, Dull HB, Eickhoff TC: An outbreak in 1965 of severe respiratory illness caused by the Legionnaires' disease bacterium. J Infect Dis 138:512, 1978.
6. Glick TH, Gregg MB, Berman B, Mallison G, Rhodes WW Jr, Kassanoff I: Pontiac fever. An epidemic of unknown etiology in a health department. I. Clinical and epidemiologic aspects Am J Epidemiol 107:149, 1978.
7. Fraser DW: Legionnaires' Disease: Four Summers' Harvest. Am J Med 68:1, 1980.
8. Bartlett CLR: Sporadic cases of Legionnaires' disease in Great Britain. Ann Intern Med 90:592, 1979.
9. Renner ED, Helms CM, Hierholzer WJ Jr, Hall N, Wong YW, Viner JP, Johnson W, Hausler WJ Jr: Legionnaires' disease in pneumonia patients in Iowa. A retrospective seroepidemiologic study, 1972–1977. Ann Intern Med 90:603, 1979.
10. Meenhorst PL, Van der Meer JWM, Borst J: Sporadic cases of Legionnaires' disease in the Netherlands. Ann Intern Med 90:529, 1979.
11. Meyer RD, Edelstein PH, Finegold SM: Legionnaires' disease in the immunocompromised host. Abstract International Symposium on Infections in the immunocompromised host. Veldhoven, The Netherlands, 1980.
12. O'H Tobin J, Beare J, Dunnill MS et al: Legionnaires' disease in a transplant unit: Isolation of the causative agent from shower baths, Lancet 2:118, 1980.
13. Dondero TJ Jr, Rentdorff RC, Mallison Gf et al: Outbreak of Legionnaires' disease associated with a contaminated air-conditioning cooling tower. N Engl J Med 302:365, 1980.

14. Fliermans CB, Cherry WB, Orrison LH, Thacker L. Isolation of Legionnaires' disease bacteria from nonepidemic aquatic habitats. Applied Environ Microbiol 37:1239, 1979.
15. Center for Disease Control. Legionnaires' disease–United States. Morbid Mortal Wkly Rep 27:439, 1978.
16. Communicable Disease Report, Legionnaires' disease 28, 1979.
17. Reid D: Benidorm episode and Legionnaires' disease. Scott Med J 23:118, 1978.
18. Grist NR, Reid D, Najera R: Legionnaires' disease and the traveller. Ann Intern Med 90:563, 1979.
19. Foy HM, Broome CV, Hayes PS et al: Legionnaires' disease in a prepared medical care group in Seattle 1963–1975. Lancet 1:767, 1979.
20. Center for Disease Control–Legionellosis–Vasterãs, Sweden, Morbid Mortal Wkly Rep 29:206, 1980.
21. Broome CV, Goings SAJ, Thacker SB, Vogt RL, Beaty HN, Fraser DW, the field investigation team. The Vermont epidemic of Legionnaires' disease. Ann Intern Med 90:573, 1979.
22. Kirby BD, Snyder KM, Meyer RD, Finegold SM: Legionnaires'disease: Report of sixty-five nosocomially axquired cases and review of the literature. Medicine 59:188, 1980.
23. Petričevič J, Presečki, Grgas J: Legionnaires' disease in two patients following renal transplantation. Abstract International Symposium on Infections in the Immunocompromised Host. Veldhoven, the Netherlands, 1980.
24. Haley CE, Cohen ML, Halter J, Meyer RD: Nosocomial legionnaires' disease: A continuing common-source epidemic at Wadsworth Medical Center. Ann Intern Med 90:583, 1979.
25. Thomason BH, Harris PP, Hicklin MD et al: A Legionella-like bacterium related to WIGA in a fatal case of pneumonia. Ann Intern Med 91:673, 1979.
26. Heber GA, Thomason BH, Harris PP et al: 'Pittsburgh Pneumonia Agent': A bacterium phenotypically similar to Legionella pneumophila and identical to the TATLOCK bacterium. Ann Intern Med 92:53, 1980.
27. Gregory DW, Schaffner W, Alford RH, Kaiser AB, McGee ZA: Sporadic cases of Legionnaires' disease: The expanding clinical spectrum. Ann Intern Med 90:518, 1979.
28. Lees AW, Tyrrell WF: Severe cerebral disturbance in Legionnaires' disease. Lancet ii: 1336, 1978.
29. Miller AC: Early clinical differentiation between Legionnaires' disease and other sporadic pneumonias. Ann Intern Med 90:526, 1979.
30. Gasper TM, Farndon PA, Davies R: Thrombocytopenia associated with Legionnaires' disease. Br Med J 2:1611, 1978.
31. Mould RF, Williams RJ: Hypoplastic anaemia associated with Legionnaires' disease. Br Med J 1:366, 1980.
32. Tsai TF, Finn DR, Plikaytis BD, McCauley W, Martin SM, Fraser DW: Legionnaires' disease: Clinical features of the epidemic in Philadelphia. Ann Intern Med 90:509, 1979.
33. Kerr DN, Brewis RA, Macrae AD: Legionnaires' disease and acute renal failure. Br Med J 2:538, 1978.
34. Case Records of the Massachusetts General Hospital: Case 17–1978. N Engl J Med 298:1014, 1978.
35. Kirby BD, Snyder KM, Meyer RD, Finegold SM: Legionnaires' disease: Clinical features of 24 cases. Ann Intern Med 89:297, 1978.
36. Friedman HM: Legionnaires' disease in nonlegionnaires: A report of 5 cases. Ann Intern Med 88:294, 1978.
37. Dietrich PA, Johnson RD, Fairbank JT, Walke JS: The chest radiograph in Legionnaires' disease. Radiology 127:577, 1978.
38. Gump DW, Frank RO, Winn WC Jr, Foster RS Jr, Broome CV, Cherry WB: Legionnaires' disease in patients with associated serious disease. Ann Intern Med 90:538.

39. Saravolatz LD, Burch KH, Fisher E, Madhaven T, Kiani D, Neblett T, Quinn EL: The compromised host and Legionnaires' disease. Ann Intern Med 90:533, 1979.
40. Hake KB, Van Dijke JJ, Gerberg E et al.: Legionnaires' disease and pulmonary cavitation. Arch Intern Med 139:485, 1979.
41. Center for Disease Control. Legionnaires' disease: Diagnosis and management. Ann Intern Med 88:363, 1978.
42. Broome CV, Cherry WB, Winn WC, MacPherson BR Jr: Rapid diagnosis of Legionnaires' disease by direct immunofluorescent staining. Ann Intern Med 90:1, 1979.
43. Edelstein PH, Meyer RD, Finegold SH: Laboratory diagnosis of Legionnaires' disease. Am Rev Resp Dis 121:317, 1980.
44. Feeley JC, Gorman GW, Weaver RE, Mackel DC, Smith HW: Primary isolation media for the Legionnaires' disease bacterium. J Clin Microbiol 8:320, 1978.
45. Feeley JC, Gibson RJ, Gorman GW, Langford NC, Rasheed JK, Mackel DC, Baine WB: CYE agar: A primary isolation medium for the Legionnaires' disease bacterium. J Clin Microbiol 10:437, 1979.
46. Edelstein PH, Meyer RD, Finegold SH: Isolation of Legionella pneumophila from blood. Lancet 1:750, 1979.
47. Macrae AD, Greaves PW: Isolation of Legionella pneumophila from blood culture. Br Med J 2:1189, 1979.
48. Thornsberry C, Kirven LA: Beta-lactamase of the Legionnaires' bacterium. Current Microbioloby 1:51, 1978.
49. Fraser DW, Wachsmuth I, Bopp C, Feeley JC, Tsai TF: Antibiotic treatment of guinea pigs infected with agent of Legionnaires' disease. Lancet 1:175, 1978.
50. Lewis VJ, Thacker WL, Shepard CC, McDade JE: In vivo susceptibility of the Legionnaires' disease bacterium to ten antimicrobial agents. Antimicrob Agents Chemother 13:–19, 1978.
51. Blackmon JA, Hicklin MD, Chandler FW, the Special Expert Pathology Panel: Legionnaires' disease: Pathologic and historical aspects of a 'new' disease. Arch pathol Lab Med 102:337, 1978.
52. Chandler FW, McDade JE, Hicklin MD, Blackmon JA, Thomason BM, Ewing EP Jr: Pathologic findings in guinea pigs inoculated intraperitoneally with the Legionnaires' disease bacterium. Ann Intern Med 90:671, 1979.
53. Horwitz MA, Silverstein SC: The Legionnaires' disease bacterium grows intracellularly in human monocytes. Abstract presented for the American Society for Clinical Investigation. Clin Res 28:512A, 1980.
54. McDade JE, Shepard CC: Virulent to avirulent conversion in the Legionnaires' disease bacterium and its effect on isolation techniques. J Infect Dis 139:707, 1979.
55. Beaty HN, Miller AA, Broome CV et Al: Legionnaires' disease in Vermont, May to October 1977. JAMA 240:127, 1978.
56. Garrod LP, Waterworth PH: Methods of testing combined antibiotic bactericidal action and the significance of the results. J Clin Path 15:328, 1962.
57. Keys TF: A sporadic case of pneumonia due to Legionnaires' disease. Mayo Clin Proc 52:657, 1977.
58. Krech U, Kohli P, Pagon S: Legionnaires' disease in Switzerland. Schweiz Med Wochenschr 108:1653, 1978.
59. Acocella G, Hamilton-Miller JMT, Brumfitt W: Can rifampicin use be safely extended? Evidence for non-emergence of resistant strains of mycobacterium tuberculosis. Lancet 1:740, 1977.
60. Meenhorst PL, Van der Meer JWM, Van Brummelen P: Een patiënt met Legionairsziekte in Nederland, Ned T Geneesk 122:507, 1978.

DISCUSSION

Dr Mattie: One of the causes of differences between erythromycin and other more potent drugs, in those retrospective studies, could of course be that only the less serious cases have been treated with erythromycin and the very severe cases have been treated with aminoglycosides. Is that supposition warranted by the publications?

Dr Meenhorst: No, it is not. Most patients were treated initially with either penicillin or a combination of penicillin, a cephalosporin or aminoglycoside. When no improvement in the clinical illness was seen, one started to think of Legionella pneumonia and erythromycin was added.

Dr Van der Waaij: In order to further reduce the number of negative sputum smears in the patients, would you advocate making a so-called half a-Gram stain in order to make these Legionella pneumonia bacteria visible. Apparently, they do not show up in the complete Gram stain and this may explain why we miss them every now and then.

Dr Meenhorst: I am responsible for the story of the half a-Gram stain also. In fact, it was performed on lung tissues. My experience now with sputa is very disappointing, because you have all kinds of material in the sputum which will color also, and then it is hard to differentiate between a bacterium and artefacts. The diagnosis can be made quickest with direct immuno-fluorescent staining of either sputum, a transtracheal aspirate or material obtained by fibroscopy.

Dr Hilvering: I do not remember your mentioning dosages. Als far as I know, you often need higher dosages of erythromycin than usual.

Dr Meenhorst: Mild cases can be treated with orally administered erythromycin. Successes have been described with a dosage of 0.5 to 1 g 4 times daily. On the other hand, in seriously ill patients where one is not certain of the absorbed amount of erythromycin, we prefer to administer 0.5 or 1 g erythromycin 4 times daily intravenously depending on the clinical situation. How long should treatment be continued? Relapses have been described after therapy was discontinued within 3 weeks. So, at this moment our advice should be: to treat for at least 3 weeks. The other question is, if treatment has to be given as long as pulmonary infiltrates are present? No data are available to support the idea that the resolution of pulmonary infiltrates can be quickned by long-term administration of erythromycin.

Dr Kerrebijn: You combined erythromycin with tobramycin intravenously? What was your indication for this combination?

Dr Meenhorst: In our hospital this is an empirical therapy in leukemic patients who are granulocytopenic and develop a lung infiltrate, have fever and produce no sputum. All patients with a proven Legionella pneumonia were thrombocytopenic, and so we were not able to do any invasive diagnostic investigations, like bronchoscopy. Erythromycin was given for Legionella pneumophila and tobramycin for possible other Gram-negative micro-organisms causing the pneumonia.

Dr Van der Laag: Have you made any observations in children, up to now? You have only shown observations in adults.

Dr Meenhorst: No, I have not. Professor Butzler from Brussels has told me that they have seen a Legionella pneumonia in a child of two years, who was treated for leukemia. It is well known by now that also immuno-compromised children seem to be at risk for Legionella pneumonia.

Dr Mattie: I can think of one other possible cause of the retrospective success of erythromycin. In those cases where erythromycin was only a second choice: it could be that patients still alive at the moment when this (second) choice was made were better off than the patient who had already died at an earlier stage of the disease.

Dr Meenhorst: I think, that is a good point you made. This may be partially true, but on the other hand, clinically, there is a rapid effect when erythromycin is given; within 24 to 48 hr the clinical condition of most patients improved. We tend to ascribe this to the antibotics given.

Dr van der Meer: I think that your strongest argument for an effective treatment comes from your leukemia patients who have zero granulocytes, severe pneumonia and recover on erythromycin together with tobramycin, although I think that tobramycin is not important in this particular situation.

Dr Meenhorst: I do not know.

18. PULMONARY INFECTIONS IN MYELOSUPPRESSED OR IMMUNOSUPPRESSED PATIENTS

J.W.M. VAN DER MEER

INTRODUCTION

In patients with defective host defense mechanisms, pulmonary infections are frequent causes of morbidity and mortality. In this review only the treatment of these infections in patients with myelosuppression and patients with immunosuppression will be discussed; disorders of humoral immunity will not be dealt with.

Since the pulmonary infections in myelosuppressed patients are different from those with immunosuppression these two categories of patients will be discussed separately. We are of course, aware that patients with both immunosuppression and myelosuppression are not infrequently encountered.

PULMONARY INFECTIONS IN MYELOSUPPRESSED PATIENTS

Granulocytopenia is the most important factor leading to infection in the myelosuppressed patient. The risk increases as the granulocyte count in the peripheral blood falls [1]. Particularly when the granulocyte count falls below $100/mm^3$, a marked increase in the number of infections is seen [2]. The infections in the myelosuppressed patients are predominantly caused by bacteria and fungi. Other types of infection, e.g. viral and protozoan, are seen when cell mediated immunity is deficient (see below). Pulmonary infections caused by bacteria or fungi are frequently encountered in granulocytopenic patients, and

Table 1. Micro-organisms causing pneumonia in the granulocytopenic patient.

Gram-negative rods (E. coli, Klebsiella spp., Pseudomonas aeruginosa)
Staphylococcus aureus
Fungi (Candida spp., Aspergillus spp., Phycomyces spp.)
Legionella pneumophila
Pneumocystis carinii
Cytomegalo virus

---

are still a leading cause of death in these patients. The most important pathogens in these patients are listed in Table 1. Most patients become infected with the microflora which has colonized their oro- and nasopharynx. Especially severely ill hospitalized patients are more prone to colonization with Gram-negative bacilli in the upper airways [3]. With increasing duration of hospital stay, more patients become colonized with Gram-negative micro-organisms in the oropharynx [3,4]. Use of antibiotics, presumably especially those that affect the colonization resistance [5] may lead to colonization with multiresistant bacteria [4] and fungi [6]. In this respect, the study of Aisen et al. [6], showing colonization with Aspergillus spp. of the nose of granulocytopenic patients treated with carbenicillin, is of importance.

Due to the shortage of granulocytes, these pulmonary infections initially show a paucity of symptoms and signs. Sickles et al. [7] found that more than one third of their patients with granulocytopenia and pneumonia had neither cough nor sputum production and about one fourth had normal physical examinations; chest X-rays indicate pneumonia in more than 95% of the cases according to these authors. However, early in the course of the infection, the chest X-rays may be normal [8]. A microbiological diagnosis is very often hampered by the lack of sputum production, and at this stage the physician is confronted with the following dilemma. Should agressive diagnostic procedures be undertaken first or should empiric ('blind') antimicrobial therapy be instituted? The diagnostic procedures which can be undertaken are listed in Table 2. In a number of patients with granulocytopenia these procedures cannot be carried out because of concomitant bleeding disorders. However, even with good hemostatic function these procedures are not without risk [9]. At present, we prefer fiberoptic bronchoscopy in these patients; to avoid contamination with oropharyngeal flora, material for microbiological investigations is taken with the help of a protected catheter (Meditech, Watertown, Mass.). The material obtained should not only be cultured, but also be examined in Gram-stained preparations. If no bacteria are seen at least the 'half-a-Gram' stain [10] and Ziehl Neelsen stain should be performed, to look for Legionella pneumophila (see chapter 17),

Table 2. Diagnostic procedures to determine the etiology of pulmonary infections.

1. Transtracheal aspiration
2. Transcricoid bronchial brush
3. Fiberoptic bronchial brush
4. Fiberoptic aspiration with a protected catheter
5. Transbronchial biopsy
6. Transpleural (thoracoscopic) biopsy
7. Needle biopsy
8. Surgical biopsy

respectively for mycobacteria, Pittsburgh pneumonia agent [11] and nocardia. The initial antimicrobial therapy should be based on these findings. A number of general rules for antimicrobial therapy in severely infected granulocytopenic patients should be followed.

1. Agents that are microbicidal are preferable. But a probably bacteriostatic drug like erytromycin proves to be effective in Legionella pneumonia in granulocytopenic patients (see chapter 17).
2. Only the parenteral route should be used for the administration of antimicrobial drugs; drugs that are normally well resorbed by the oral route, may show impaired resorption in hospitalized patients [12, 13].
3. The drugs should be administered in high dosages. Experiments in animals suggest that in neutropenic animals and in animals with impaired phagocytosis the dosage should be higher than in normal animals to obtain an adequate effect [14, 15, 16].
4. Aminoglycosides should not be given as single agents in granulocytopenic patients, since both clinical studies [16] and animal experiments [17,18] suggest that these agents are not very effective in granulocytopenia. Unpublished results of experiments carried out by Van der Voet and Mattie suggest that it might be merely the narrow therapeutic index that renders these drugs relatively ineffective in the absence of granulocytes. An effect equal to that in non-neutropenic control animals can only be obtained with much higher dosages of these drugs.

Administration of aminoglycosides by continuous infusions in granulocytopenic patients has been claimed to be more effective than by intravenous injections [19]. However, these data have not been confirmed by others, and the occurrence of toxicity (especially ototoxicity), which might be expected to be more common, with continuous infusions is a major point of concern.

In our opinion, the choice of the antimicrobial agents is also determined by the results of surveillance cultures, by what is known from the hospital flora and its sensitivity patterns, and by previous infections and antimicrobial therapy in that particular patient [20]. The value of surveillance cultures (cultures taken from nose, throat, skin, stools and urine at regular intervals during granulocytopenia) is disputed; however, the predictive value of such an inventory has at least been demonstrated for Pseudomonas aeruginosa [21, 22, 23] and for Aspergillus spp. [6].

Studies have been carried out to determine the best antibiotic regimen for initial treatment in the granulocytopenic patient [19, 24, 24, 26]. In our view, a number of problems are connected with the aim and the realization of these studies. Such regimens are not patient tailored; that means, a number of the above mentioned parameters in the choice of the antimicrobials are not taken into account and a confectionary therapy, a general scheme for all patients, is given. Thus the antimicrobial spectrum of the applied antimicrobial drugs may be unnecessarily wide, with serious consequences for the colonizing microflora and colonization resistance (see Chapter 5). In general, the patient will be colonized by micro-organisms resistant to the antimicrobial drugs applied, and the next infection in the patient may be caused by these micro-organisms. This situation can lead to serious problems for further therapy. Moreover, in most of these studies fever alone was the sole indication on which antimicrobial therapy was started immediately, but this practice minimises the possibilities for thorough microbiological investigation, which is essential for optimal therapy. The antimicrobial drugs, which are currently the most appropriate against the various micro-organisms are listed in Table 3. Also a number of drugs, which hold a promise for the future, but need more study before their use can be recommended, are included.

It should be mentioned that the efficacy of the antifungal drug miconazole has been quite convincingly shown in a large number of clinical studies [27]. However, the comparison of this drug with amphotericin B (the latter in combination with 5 fluorocytosin) has not yet been made in carefully conducted clinical trials. Nevertheless, in view of its efficacy and the relative lack of toxicity in comparison with amphotericin B, we have given miconazole a place in our armentarium.

If antimicrobial therapy is not successful in the granulocytopenic patient, suppletion of granulocytes seems to be a logical step. Although the efficacy of granulocyte transfusions has been shown in both animals and humans [28, 29,

Table 3. The cause of pulmonary infection and recommended treatment in granulocytopenic patients.

Causative organism	Therapy	
	At present	In the future
E. coli	cefamandol	cefotaxim/moxylactam
Klebsiella spp.	+	+
Proteus spp.	gentamicin	gentamicin
Ps. Aeruginosa	ticarcillin	azlocillin/piperacillin/cefaperazone
	+	+
	tobramycin/amikacin	tobramycin/amikacin
Staph. aureus	cloxacillin	
Legionella pneumophila	erytromycin	
	(rifampicin)	
Candida spp.	miconazole	ketoconazole
	or amphotericin B	
	+ 5 FC	
Aspergillus spp.	amphotericin B	econazole

30] the indications are by no means settled. It should be kept in mind that the number of granulocytes one can transfuse is small in comparison with the numbers of granulocytes present at an inflammatory site in a nongranulocytopenic patient or even during moderate granulocytopenia. Moreover, these transfusions put a heavy burden on bloodbank and donors, are accompanied by high costs[31] and are not without side effects for the recipient as has been discussed elsewhere[32, 33]. The side effects include: chills and fever following the transfusion; sensitization of the acceptor for alloantigens (this may hamper future transfusions with platelets); accidental stem cell grafting with subsequent graft-versus-host disease (preventable by irradiation of the leucocyte suspension); transmission of infectious agents such as hepatitis B virus, but especially cytomegalovirus (in a recent series a high proportion of the patients receiving prophylactic granulocyte transfusions were reported to develop cytomegalovirus infection [33]). For all these reasons we think that granulocyte transfusions are only indicated in patients with less than 100 granulocytes/mm^3 and a severe bacterial infection (e.g. pneumonia) that is not responding or likely to respond to antibiotic treatment[32].

The prevention of severe infection in the granulocytopenic patients will only be dealt with very briefly here (for reviews see refs. [34, 35, 36]). Pulmonary infections have been shown to be prevented to some extent by protected isolation of the patient [37]. Decontamination procedures seem to further reduce the number of infections in granulocytopenic patients [37, 38, 39]. Partial antibiotic decontamination [40] aims to prevent infections by elimination of the potential pathogenic aerobic micro-organisms of the patient by means of orally administered antimicrobial agents. These agents are chosen so that they do not affect the anaerobes, thereby leaving the colonization resistance intact (for detailed discussion on colonization resistance (see chapter 5)). Several investigators now use cotrimoxazole for decontamination [41]. However, a number of objections can be made against use of cotrimoxazole for decontamination. Unlike the drugs used in the partial decontamination regimen, cotrimoxazole is very valuable for treatment of infections and might become less so by widespread prophylactic use. Since cotrimoxazole is not active against Pseudomonas aeruginosa, it will not eliminate this micro-organism from the gut; in contrast, since the partial antibiotic decontamination regimen includes polymyxin B it eliminates Pseudomonas aeruginosa. Moreover, when cotrimoxazol is used, one should be aware of the immunological and non-immunological effects of the components of cotrimoxazole (sulfamethoxazole and trimethoprim) on blood cells (even transfused ones) in these (haematological) patients [42, 43].

PULMONARY INFECTIONS IN IMMUNOSUPPRESSED PATIENTS

In the immunosuppressed patient there is a disorder of cell mediated immunity. In other words, the functioning of T lymphocytes and/or of macrophages is disturbed. Such disorders may be brought about by lymphoproliferative diseases, neoplasms and by immunosuppressive drugs (e.g. glucocorticosteroids, cyclophosphamide, azathioprine). Most micro-organisms that cause infections in these patients are the so-called facultative and obligate intracellular micro-organisms; in the normal host, macrophages activated by T products from lymphocytes cope with these micro-organisms. However, not all intracellular micro-organisms can be eleminated by the normal host, and these persisting micro-organisms may give rise to endogenous reinfection, when cellular immunity becomes deficient. The immunocompromised host is of course also more susceptible to exogenous micro-organisms. The most important micro-organisms that cause pulmonary infection in the immunosuppressed patient are listed in Table 4.

In general, the clinical picture is insidious, and progressive dyspnoea is prominent. The chest X-rays show diffuse infiltrates and blood gas analysis show a fall in arterial oxygen tension. The differential diagnosis includes many micro-organisms and also non-infectious causes (effects of drugs, pulmonary involvement of the underlying disease, etc.). In order to institute proper therapy as soon as possible it is essential that diagnostic procedures are undertaken without delay (see Table 2). Since the abnormalities are often located in the interstitial space of the lung, biopsies taken with the fiberoptic bronchoscope will often be negative. Such negative findings lead to delay in diagnosis and therapy in these often rapidly progressive pulmonary conditions. We are currently investigating the

Table 4. Cause of pulmonary infections and their treatment in patients with impaired cell-mediated immunity.

Cytomegalovirus	? immune plasma
Herpes simplex	acyclovir, adenine arabinoside
Varicella zoster virus	acyclovir, adenine arabinoside
Mycobacteria spp.	antimycobacterial drugs
Nocardia spp.	cotrimoxazole; ampicillin + erytromycin
Legionella pneumophila	erytromycin (+ rifampicin)
Pittsburgh pneumonia agent	erytromycin
Chlamydia trachomatis	erytromycin
Cryptococcus neoformans	amphotericin B + 5 Fluorocytosine (miconazole, ketoconazole)
Aspergillus spp.	amphotericin B (econazole?)
Pneumocystis carinii	cotrimoxazole (pentamidine)
Toxoplasma gondii	daraprim + sulphonamide

efficacy of transpleural thoracoscopic biopsies in these patients. The most important drugs for the various pulmonary infections are listed in Table 4.

For cytomegalovirus an effective therapy is not available. The effect of immune plasma is uncertain and not without danger [44]. For herpes simplex and varicella zoster virus, Aciclovir seems to be least toxic effective therapy [45].

In the immunosuppressed patient mycobacterium tuberculosis, but also the atypical mycobacteria [46], may be the cause of severe infection. For the latter group of mycobacteria, the sensitivity for the first line antimycobacterial drugs (isoniazide, rifampicin and ethambutol) cannot be predicted and laboratory data must be obtained for the institution of optimal therapy. Of course a provisional therapy with antimycobacterial drugs should be given a soon as possible. For the treatment of Nocardia spp. infections a large number of drugs are mentioned in the literature. Probably cotrimoxazole is effective [47]. Synergism has been reported for erytromycin and ampicillin [48]. Legionella pneumonia is discussed extensively in Chapter 17. Erytromycin seems to be effective in infections caused by Pittsburgh pneumonia agent [11] and Chlamydia trachomatis pneumonia [49]. The latter, which has originally been described in infants, has recently been described in adult immunosuppressed patients [50]. Cryptococcal lung disease in its extensive forms should probably be treated with amphotericin B and 5 fluorocytosin. This combination has been shown to be synergistic both in vitro and in vivo [51, 52, 53]. Miconazole, and perhaps the oral drug ketoconazole, may be effective in this condition. For aspergilllus infections amphotericin B is still the drug of choice. However, econazole, which shows a high activity against Apergillus spp. in vitro, might be effective in aspergillus infections. Cotrimoxazole given in high dosage (20 mg trimethoprim and 100 mg sulfamethoxazole/kg/day) has been shown to be very effective in Pneumocystis carinii pneumonitis [54]. The more toxic drug pentamidine is seldom used nowadays. For toxoplasma infections, daraprim together with a sulfonamide, is the only treatment with proven efficacy [55].

CONCLUDING REMARKS

Nowadays a large number of highly effective drugs are available for the treatment of pulmonary infections in the myelosuppressed and/or immunosuppressed patient. Since host defence mechanisms are defective in these patients, selection of the most effective treatment is of utmost importance. However, it is not possible to select the optimal therapy for the individual compromised patient without a proper microbiological diagnosis.

REFERENCES

1. Bodey GP, Buckley M, Sathe YS, Freireich EJ: Quantitative relationship between circulating leucocytes and infection in patients with acute leukemia. Ann Int Med 64:328, 1966.
2. Van der Meer JWM, Boekhout M, Alleman M: Infectious episodes in severely granulocytopenic patients. Infection 7:171, 1979.
3. Johanson WG, Pierce AK, Sanford JP: Changing pharyngeal bacterial flora of hospitalized patients. Emergence of Gram-negative bacilli. New Engl J Med 281:1137, 1969.
4. Pollack M, Charache P, Nieman RE, Jett MP, Reinhardt JA, Hardy PH: Factors influencing colonisation and antibiotic-resistance patterns in Gram-negative bacteria hospital patients. Lancet 2:688, 1972.
5. Van der Waay D, De Vries JM, Lekkerkerk JEC: Colonisation resistance of the digestive tract in conventional and antibiotic treated mice. J Hyg (Lond) 69:405, 1971.
6. Aisner J, Murillo J, Schimpff SC, Steere AC: Invasive aspergillosis in acute leukemia correlation with nose culture and antibiotic use. Ann Int Med 90:4, 1979.
7. Sickles EA, Young VM, Greene WH, Wiernik PH: Pneumonia in acute leukemia. Ann Int Med 79: 528, 1973.
8. Schimpff SC, Young VM, Greene WH, Vermeulen GD, Moody MR, Wiernik PH: Origin of infection in acute nonlymphocytic leukemia. Significance of hospital acquisition of potential pathogens. Ann Int Med 77:707, 1972.
9. Ries K, Levison Me, Kaye D: Transtracheal aspiration in pulmonary infection, Arch Int Med 133:453, 1974.
10. De Freitas JL, Borst J, Meenhorst PL: Easy visualisation of Legionella pneumophila by 'half-a-Gram' stain procedure. Lancet I:270, 1979.
11. Myerowits RL, Pasculle AW, Dowling JN, Pazin GJ, Puerzer M, Yee RB, Rinaldo Jr CR, Hakala Tr: Opportunistic lung infection due to 'Pittsburgh pneumonia agent'. New Eng J Med 301:953, 1979.
12. Kunst MW, Mattie H: Absorption of pivampicillin in postoperative patients. Antimicrob. Ag Chemother 8:11, 1975.
13. Mattie H, Meenhorst PL: Clinical pharmacological evaluation of aminopenicillin. Infection, suppl 5:456, 1979.
14. Kunst MW, Mattie H, Van Furth R: Antibacterial efficacy of cefazolin and cephradine in neutropenic mice. Infection 7:30, 1979.
15. Bakker-Woudenberg IAJM, Van Gerven ALEM, Michel MF: Efficacy of antimicrobial therapy in experimental rat pneumonia: antibiotic treatment schedules in rats with impaired phagocytosis. Infect Immun 25:376, 1979.
16. Bodey GP, Middleman E, Umsawadi T, Rodriquez V: Infections in cancer patients. Results with gentamicin sulfate therapy. Cancer 29:1697, 1972.
17. Dale DC, Reynolds HY, Pennington JE, Elin RJ, Herzig GP: Experimental Pseudomonas pneumonia in leukopenic dogs: comparison of therapy with antibiotics and granulocyte transfusions. Blood 47:869, 1976.
18. Mattie H, Van der Voet B: The relative potency of amoxicillin and ampicillin in vitro and in vivo. Scand J Infect Dis. Accepted for publication.
19. Keating MJ, Bodey GP, Valdivieso M: A randomized comparative trial of three aminoglycosides–Comparison of continuous infusions of gentamicin, amikacin and sisomicin combined with carbenicillin in the treatment of infections in neutropenic patients with malignancies. Medicine (Baltimore) 58:159, 1979.
20. Nauta EH, Van Furth R: Infection in immunosuppressed patients. Infection 4:202, 1975.
21. Schimpff SC, Moody M, Young VM: Relationship of colonization with Pseudomonas aeruginosa to development of pseudomonas bacteremia in cancer patients. Antimicrob Ag Chemother Jaarboek 240, 1970.

22. Bodey GP, Rodriquez V: Advances in the management of Pseudomonas aeruginosa infections in cancer patients. Europ J Cancer 9:435, 1973.

23. Meyer DV, Winston D, Young LS, Gale RP, Martin WJ: Surveillance cultures in immunosuppressed patients: what do they mean? In: Current chemotherapy and infectious disease. Proceedings of the 11th international congress of chemotherapy and the 19th interscience conference on antimicrobial agents and chemotherapy (Nelson JD, Grassi C, eds.), Washington DC, American Soc. for Microbiology, 1980, p. 1436.

24. Schimpff SC, Satterlee W, Young ME, Serpick A: Empiric therapy with carbenicillin and gentamicin for febrile patients with cancer and granulocytopenia. New Engl J Med 284:1061, 1971.

25. Tattersall MHN, Spiers ASD, Darrell JH: Initial therapy with combinations of five antibiotics in febrile patients with leukaemia and neutropenia. Lancet I:162, 1972.

26. EORTC international antimicrobial therapy project group. Three antibiotic regimens in the treatment of infection in febrile granulocytopenic patients with cancer. J Infect Dis 137:14, 1978.

27. Medoff G, Kobayashi GS: Strategies in the treatment of systemic fungal infections. New Engl J Med 302:145, 1980.

28. Higby DJ, Yates JW, Hendurson ES, Holland JF: Filtration leukopheresis for granulocyte transfusion therapy: clinical and laboratory studies. New Engl J Med 292:761, 1975.

29. Hertzig RH, Hertzig GP, Graw RG, Bull MI, Rai KK: Successful granulocyte transfusion therapy for Gramnegative septicemia. New Engl J Med 296:701, 1977.

30. Alavi JB, Root RK, Djerassi I, Evans AE, Gluckmann SJ, McGregor RR, Dupont G, Schryber AD, Schaw JM, Koch P. Cooper RA: A randomized clinical trial of granulocyte transfusion for infection in acute leukemia. New Engl J Med. 296:706, 1977.

31. Rosenshein MS, Farewell VT, Price TH, Larson EB, Dale DC: The cost effectiveness of therapeutic and prophylactic leukocyte transfusion. New Engl J Med 302:1058, 1980.

32. Eernisse JG, Van der Meer JWM: Granulocyte transfusions. Neth J Med 22:128, 1979.

33. Young LS: Leucocyte transfusions In: Current chemotherapy and infectious disease. Proceedings of the 11th international congress of chemotherapy and the 19th interscience conference on antimicrobial agents and chemotherapy (Nelson JD, Grassi C eds.), Washington DC, American Soc. for Microbiology, 1980, p. 1431.

34. Guiot HFL: Prevention of infection and partial antibiotic decontamination. Neth J Med 22:114, 1979.

35. Van der Waay D: Protective isolation for prevention of infection in patients with severly decreased resistance. Neth J Med 22: 112, 1979.

36. Editorial. Infection complicating severe granulopenia. Lancet 1:25, 1980.

37. EORTC gnotobiotic project group, writing committee: Protective isolation and antimicrobial decontamination in patients with high susceptibility to infection I. Clinical results. Infection 5:107, 1977.

38. Levine AS, Siegel SE, Schreiber AD, Hauser J. Preisler HD, Goldstein IM, Seider F, Simon R, Perry S, Bennet JE, Henderson ES: Protected environments and prophylactic antibiotics. A prospective controlled study of their utility in the therapy of acute leukemia. New Engl J Med 288:477, 1973.

39. Schimpff SC, Greene WH, Young VM, Fortner CL, Jepsen L, Cusack N, Block JB, Wiernik PH: Infection prevention in acute nonlymphocytic leukemia. Ann Int Med 82:351, 1975.

40. Guiot HFL, Van Furth R: Partial antibiotic decontamination. Brit Med J 1:800, 1977.

41. Gurwith MJ, Brunton JL, Lash B: A prospective controlled investigation of prophylactic trimethoprim/sulphamethoxazole in hospitalized granulocytopenic patients. Am J Med 66:248, 1979.

42. Claas FHJ, Van der Meer JWM, Langerak J: Immunological effect of cotrimoxazole on platelets. Brit Med J 2:898, 1979.

43. Frisch JM: Clinical experience with adverse reactions to trimethoprim sulphamethoxazole. J Infect Dis (Suppl) 128:607, 1973.

44. Dijkmans BAC, Versteeg J, Kauffmann RH, Van den Broek PJ, Eernisse JE, Van Zanten JJ, Bakker W, Kalff MW, Van Hooff JP: Treatment of cytomegalovirus pneumonitis with hyperimmune plasma, Lancet I:820, 1979.

45. Selby PJ, Powles RL, Jameson B, Kay HEM, Watson JG, Thornton R, Morgenstern G, Clink HM, McElwain TJ, Prentice HG, Corringham R, Ross MG, Hoffbrand AW, Brigden D: Parenteral acyclovir therapy for herbes virus infections in man. Lancet 2:1267, 1979.

46. Wolinsky E: Nontuberculous mycobacteria and associated diseases. Am Review Resp Dis 119:107, 1979.

47. Maderazo EG, Quintiliani R: Treatment of nocardial infection with trimethoprim and sulfamethoxazole. AM J Med 57:671, 1974.

48. Bach MC, Monaco AP, Finland M: Pulmonary nocardiosis therapy with minocycline and with erythromycin plus ampicillin. JAMA 224:1378, 1973.

49. Harrison HR, English MG, Lee CK, Alexander ER: Chlamydia trachomatis infant pneumonitis. Comparison with matched controls and other infant pneumonitis. New Engl J Med 298:702, 1978.

50. Tack KJ, Peterson PK, Rasp FL, O'leary M, Hanto D, Simmons RL, Sabbath LD: Isolation of Chamydia trachomatis from the lower respiratory tract of adults. Lancet I:116, 1980.

51. Medoff G, Comfort M, Kobayashi GS: Synergistic action of amphotericin B and 5-fluorocytosine against yeastlike organisms. Proc Soc Exp Biol Med 138:571, 1971.

52. Block ER, Bennet JE: The combined effect of 5-fluorcytosine and amphotericin B in the therapy of murine cryptococcosis. Proc Soc Exp Biol Med 142:476, 1976, 1973.

53. Bennett JE, Dismukes WE, Duma RJ, Medoff G, Sande MA, Gallis H, Leonard J, Fields BT, Bradshaw M, Haywood H, McGee ZA, Cate TR, Cobbs CG, Warner JF, Alling DW: A comparison of amphotericin B alone and combined with flucytosine in the treatment of cryptococcal meningitis. New Engl J Med 301:126, 1979.

54. Hughes WT, Feldman S, Sanyal SK: Treatment of Pneumocystis carinii pneumonitis with trimethoprim-sulfamethoxazole. Canadian Med Ass J 112:47, 1975.

55. Krick JA, Remington JS: Toxoplasmosis in the adult – an overview. New Engl J Med 298:550, 1978.

DISCUSSION

Dr Mattie: I would like to come back to the advice to use high dosages of antibiotics in agranulocytic patients and the advice not to treat with gentamycin or other aminoglucosides alone. In fact, these may come to the same thing, because, if one would double the dose of aminoglycosides, this would improve results greatly. But, of course, no doctor would do that. So, probably there is nothing peculiar about aminoglycosides in treating this kind of patients, it is just that we do not want to give higher dosages of these toxic antibiotics. Therefore, the only way to improve the antibiotic effect is adding a second antibiotic. Do you agree?

Dr Van der Meer: I agree, but I think you have some data on animal experiments which are not completely in line with this view. Perhaps you could comment on that.

Dr Mattie: To a certain extent, they are in line. They show that if a higher dosage of an aminoglycoside is given in agranulocytic animals with an infection, one gets the same results as with lower doses in normal infected animals. There is nothing peculiar about that. So, what I am much against is the often-heard statement that aminoglycosides are not able to counter infection in agranulocytic patients. Nobody ever tried it by giving high dosages.

I wonder whether this relative in adequacy of aminoglycosides applies only to infections where the sensivity of the micro-organism is marginal, or whether it also applies to infections by very sensitive micro-organisms, so that in fact the ordinary dosage of aminoglycosides would be relatively high.

Dr Van der Meer: Most studies in experimental animals have been carried out with Pseudomonas aeruginosa, a rather insensitive micro-organism so I cannot answer this particular part of the question. I am not sure that the results of therapy in the clinical studies with aminoglycosides in granulocytopenic patients show much difference for relatively insensitive organisms (P. aeruginosa) and more sensitive bacteria.

Dr Mouton: Perhaps I can add that, according to my experience, when these patients are treated with aminoglycosides alone, you can have positive blood cultures during therapy, even in the case of rather low MIC values. For aminoglycosides it is a little different than for other antibiotics.

Dr Gould: One further observation that we have made from time to time in such patients, although it is not confined to immunosuppressed patients: when an aminoglycoside is being used alone, and often in high dosages, the selection of Gram-positive cocci, particularly Streptococcus pneumoniae, occurs and this supersedes as a continuing infection in these patients.

Chairman: I would like to ask you about the toxoplasma. Do you think a toxoplasma infection can be treated with high doses of co-trimoxazole? There are some animal experiments that indicate a beneficial effect. Or do you prefer, especially in the immunocompromises patients, the combination of a sulphonamide and pyrimethamine?

Dr Van der Meer: The data are rather conflicting. Some people have shown an effect of co-trimoxazole in experimentally infected mice, but others cannot demonstrate such an effect. I do not know of any good clinical studies in toxoplasmosis. Thus I would recommend a sulphonamide with pyrimethamine.

Dr Hilvering: You did not discuss patients with agammaglobulinemia. I would like to know whether you advise to give these patients antibiotics continuously and whether you give them gammaglobulin routinely?

Dr Van der Meer: We think there is no indication for continuous antibiotic treatment. In general adults with agammaglobulinemia receive 50 mg gammaglobulin per kg body weight weekly by giving about 10 ml of human plasma per kg body weight per two weeks intravenously. With that regimen, you can maintain immunoglobulin levels higher than 200 mg% which has been shown to be protective in trials in Great Britain and the U.S.

Dr Weemaes: Do your patients with agammaglobulinemia have bronchiectasis?

Dr Van der Meer: The patients we currently have do not have bronchiectasis. I have limited experience in the chronic support of patients with bronchiectasis. I have seen more problems in patients with bronchiectasis on long-term antibiotics than in those without.

Dr Van der Laag: I would like to add that in our pediatric practice, we advocate more or less the same regimen of substitution therapy in agammaglobulinemic patients. Just recently, we encountered one patient, who had been on gammaglobulin substitution for a very long time, who developed pseudomonas infection in his lungs which was not cleared with aminoglycosides and azlocillin. He has very extensive bronchiectasis, for which lobectomy has now been performed. It is a matter of speculation whether such a situation will occur again, but, normally only antibiotics are indicated in cases of severe infection.

Dr Van der Meer: Pseudomonas aeruginosa is not a typical pathogen for agammaglobulinemic patients. However any patient with anatomic abnormalities of this airways may get colonized by this micro-organism, mainly as a consequence of previous course of antibiotic therapy.

19. PREVENTION OF RESPIRATORY INFECTIONS BY VACCINATION

R. VAN FURTH

PRINCIPLES OF VACCINATION

Vaccination can be defined as the induction of protective immunity by the application of living or dead micro-organisms or their products. Vaccines against various bacterial and viral pathogens are available and research in respect of certain protozoal infections is encouraging, but it will be a long time before we have vaccines against fungal infections and multicellular micro-organisms.

A vaccine is an antigen that induces a specific immune response. Such antigens include killed micro-organisms, attenuated living micro-organisms with very low virulence, antigenic subunits of viruses, cell-wall antigens, ribosomes or ribosomal extracts of bacteria, lipopolysaccharides, and modified exotoxins with reduced toxicity.

The specific immune response evoked by vaccination may be a humoral and/or cellular reaction. Under humoral antibody response is understood an increase of antibody production, the site depending on the route of vaccination. Parenteral application usually leads only to an increase of serum antibodies. This means that sufficient numbers of antibodies may be present in the circulation and to a lesser extent in the tissues, depending on the immunoglobulin class of the antibody (IgM antibodies remain in the circulation, whereas IgG antibodies may diffuse into tissues). However, at the mucosal membranes, which are often the port of entry for pathogenic micro-organisms, the antibody level may be low, except when antibodies are secreted, for example in saliva. The antibody response after parenteral vaccination follows generally a fixed pattern: at first, mainly IgM antibodies are formed; later, especially after repeated vaccination, predominantly IgG antibodies (Fig. 1). The production of IgG antibodies by B lymphocytes and plasma cells, which occurs mainly in the bone marrow but also in the lymph nodes and spleen, provides protection that may last several years. These antibodies have various kinds of activity, e.g., microbicidal, opsonic, neutralizing. It is still not yet clear whether these effects are brought about by the same or different antibody (immunoglobulin) molecules. Different patterns of immunoglobulins are seen in the mucosal membranes and the serum. At the

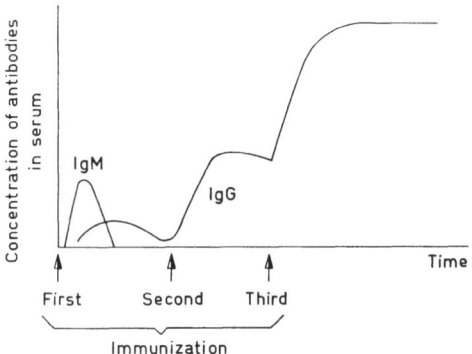

Fig. 1. Course of immunoglobulin classes of antibodies after repeated parenteral immunization.

former site the major immunoglobulin is secretory IgA (IgA-Sc), although IgG and IgM antibodies are also present. IgA-Sc is synthesized locally, the IgA by the plasma cells in the mucosa and the secretory component by epithelial cells. Once produced, these two proteins are secreted and then combined to form a single molecule, IgA-Sc (Fig. 2). IgG and IgM antibodies, too, are synthesized by lymphoid and plasma cells localised in the mucosa. Secretory IgA differs from IgG, IgM, and serum IgA in many characteristics. The functions of IgA-Sc are summarized in Table 1. It is clear that these antibodies can be very effective in the defence against penetrating micro-organisms, especially in combination with other host factors such as the ciliary action and mucus secretion in the respiratory tract, gastric acid secretion and bowel movement of the gastro-intestinal tract, and peristalsis of the ureter as well as adequate urine drainage from the bladder for the uropoietic tract.

However, optimal antibody response of the mucosal membranes, i.e., the local production of IgA-Sc, IgM, or IgG antibodies, requires local antigen stimulation. Parenteral antigenic stimulation usually does not lead to the formation of antibodies in the mucosa, because antigens do not reach those tissues after

Table 1. Functions of secretory IgA.

No bacterial activity except when complement and lysozyme are present
No opsonization of micro-organisms
Complement activation by alternative pathway; not by classical pathway
Prevention of adhesion of bacteria to mucosal surface
Inhibition of motility of bacteria
Agglutination of bacteria
Inhibition of antigen uptake by mucosa
Neutralization of enterotoxins
Virus neutralization

240

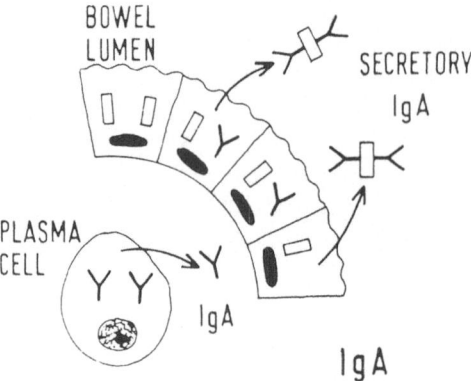

Fig. 2. Synthesis of IgA in mucosal membranes.

parenteral administration. Effective application for optimal IgA-Sc antibody formation could be accomplished by aerosol or orally, but unfortunately, most antigens in vaccines are not in a form permitting use of these routes.

An increase of cell-mediated immunity involves a different principle. The best example is vaccination with BCG, which is based on the classical experiments of Paul Römer (1909)[1]. Koch showed that when normal guinea pigs are inoculated subcutaneously with a large number of live tubercle bacilli, the wound closes and seemingly heals, but after 10 to 14 days the animal develops extensive ulceration which is ultimately fatal. However, when a small number of tubercle bacilli are applied intracutaneously, the animal's defences can cope with them and it will survive. After such animals recover, their immunological status is changed in that a state of immunity has developed; if they are re-infected with a large number of live tubercle bacilli 4 to 6 weeks later, they will recover. Meril Chase [2, 3] demonstrated in 1942 that this kind of immunity can be transferred from an immune to a non-immune guinea pig by lymphocytes. The mechanism underlying this form of immunity, called cell-mediated immunity, is now known (Fig. 3). During the first contact of an antigen with T lymphocytes, sensitization occurs, i.e., T lymphocytes with receptors for that particular antigen proliferate. When a second contact occurs between the specific antigens and the sensitized lymphocytes, the T lymphocytes produce a lymphokine. This substance activates macrophages, which then ingest, kill, and digest micro-organisms more effectively; however, the increased bactericidal activity of cells is not antigen specific; other micro-organisms too can be dealt with more effectively.

KIND OF VACCINATION

The purpose of vaccination is to increase host defence against a certain micro-organism. This can be achieved by stimulation of antibody production and/or an

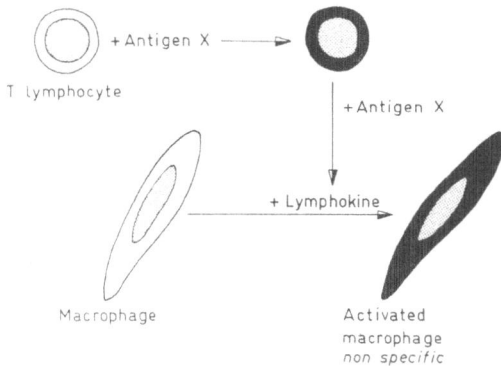

Fig. 3. Schematic representation of cell-mediated immunity, showing the specific interaction between T lymphocytes and antigen as well as the non-specific activation of macrophages.

increase of cell-mediated immunity. The nature and intensity of the immune reaction induced by vaccination depend on such factors as the characteristics of the antigen, the interval after previous vaccination, the age and condition of the individual. For practical purposes, I distinguish three types of vaccination: (a) routine, (b) selective, and (c) elective [4].

Routine vaccination comprises vaccination of all children and adults according to rules set by the government.

Selective vaccination includes vaccination of certain groups of healthy people who run an increased risk of becoming infected with certain micro-organisms.

Elective vaccination is the vaccination of patients to increase resistance to certain micro-organisms involving an increased risk.

Routine vaccination includes vaccination against diphtheria, poliomyelitis, tetanus, whooping cough, and measles; in the future vaccination against rubella and mumps will perhaps be added. Examples of selective vaccination are vaccination of medical and military personnel and employees of large concerns against influenza, vaccination against yellow fever and cholera for travellers to an endemic area, and rabies vaccination after a bite by an infected animal. A classical example of elective vaccination is the vaccination of special categories of patients against influenza (see below), pneumococcal vaccination of certain kinds of patients (see below), and pseudomonas vaccination of burn patients. The routes for application of vaccines are intradermal, subcutaneous, intramuscular, or respiratory (by aerosol).

VACCINATION FOR RESPIRATORY INFECTIONS

Vaccines which are or will soon become available for immunization are listed in Table 2.

Table 2. Vaccines for prevention of respiratory infections.

Disease or micro-organism	Type of vaccine*	Kind of application
Diphtheria	toxoid	routine
Whooping cough	inactivated	routine
Measles	attenuated	routine
Influenza	inactivated attenuated or split	selective and elective
Adenovirus**	inactivated or attenuated	selective and elective
Respiratory-syncytial virus**	inactivated or attenuated	selective and elective
BCG	attenuated	selective
Strep. pneumoniae	polysaccharide	selective and elective
B-haemolytic streptococci**	M protein	selective and elective
H. influenzae**	polysaccharide	selective and elective
Cytomegalovirus**	attenuated	selective and elective

* toxoid = modified exotoxins with reduced toxicity; inactivated = killed micro-organisms; attenuated = attenuated living micro-organisms with very low virulence; polysaccharide = capsular polysaccharides; split = antigenic subunits of the virus; M protein = cell wall antigen.
** not yet available.

Since the introduction of vaccination of children before the age of one year against diphtheria and whooping cough and since 1976 against measles, the incidence of these infections has dropped to almost nil. In The Netherlands, compliance for vaccination was in 1978 93.4% of the infants younger than 14 months, and for revaccination at 4 and 9 yr also about 94%. Selective vaccination with influenza virus is indicated for medical personnel (physicians, nurses, physiotherapists, laboratory personnel, hospital transport workers, and office staff, etc.), public-service employees (sanitation department, health department, police, etc.), military personnel, and employees of large concerns. Elective vaccination is considered for patients with a chronic heart disease, chronic respiratory disease, metabolic disease (e.g., diabetes mellitus), kyphoscoliosis, or mucoviscidosis, or older persons, and for pregnant women, the last category only when threatened by an epidemic or pandemic.

Influenza vaccine, either an attenuated live vaccine or a split vaccine consisting of small antigenic subunits of the virus surface, is effective in the prevention of infection with influenza A and B virus, although not in all vaccinated individuals. In contrast with what was expected it has been shown that patients with a lymphoproliferative disease and patients with kidney transplants, both receiving immunosuppressive therapy, produce antibody on a level that can be considered to give protective immunity.

Pneumococcal vaccine is a relatively new polysaccharide vaccine containing antigens of 14 serotypes of pneumococci. It has been shown that this polyvalent vaccine induces the production of antibodies that persist for as long as five years. Asplenic patients and patients with sickle-cell anaemia are more prone to life-threatening pneumococcal infections and should be vaccinated. In patients who are to undergo elective splenectomy (e.g., for spherocytosis), vaccination before surgery would be recommended; vaccination after splenectomy may also be effective. Patients with altered immune response due to agammaglobulinaemia, lymphoproliferative diseases, immunosuppressive drugs or radiotherapy, may not respond optimally to the vaccine. Older patients, individuals with congestive heart failure, chronic pulmonary disease, (alcoholic) cirrhosis, or diabetes mellitus, and children with the nephrotic syndrome for whom pneumococcal infections are more dangerous, must also be considered candidates for pneumococcal vaccination. The vaccination must be repeated after three years but not before that, to avoid an Arthus reaction at the injection site due to interaction between the pneumococcus antigen and persisting circulating antibodies.

BCG vaccine has proved to be effective against infections with M. tuberculosis but involves some risk in patients with a decreased immune response, because in such patients this live vaccine, which is safe in healthy individuals, may cause a disseminated infection. This possibility should be taken into account when BCG is used in immunotherapy.

Cytomegalovirus causes life-threatening infections in patients with impaired immune responses. For most of these patients it is not known whether the cause is a recent infection with cytomegalovirus or reactivation of a latent infection persisting in the lymphocytes. A live vaccine against cytomegalovirus is under development.

A vaccine against respiratory-syncytial virus has been used, but often led to complications in the form of pneumonitis. This was first thought to be an Arthus reaction due to interaction between circulating (maternal) antibodies and the antigen of the vaccine. Another possibility, recently shown to occur for myxoviruses, is that antibodies may enhance the growth of (attenuated) virus in cells such as macrophages and thus cause an inflammation. In connection with the development of vaccines for encapsulated H. influenza it is of importance that some capsule antigens of H. influenza cross-react with E. coli.

PASSIVE PROTECTION

Administration of immunoglobulins in the form of 16% gammaglobulin or hyperimmune serum can be effective in the prevention or treatment of some kinds of infectious diseases. Hyperimmune serum for a specific respiratory infection is not available except for diphtheria exotoxins. Gammaglobulin preparations, which are made from pooled donor plasma, contain a variety of antibodies, albeit in low concentrations. In patients with congenital or acquired agammaglobulinaemia such preparations may prevent respiratory infections if given regularly in an adequate dosage (1 ml 16% gammaglobulin per kg body weight every 3 weeks), but it is usually impossible to continue the intramuscular administration of gammaglobulin for a period of years. Most of the intravenous gammaglobulin preparations are treated with a proteolytic enzyme (e.g. plasmin or pepsin) in order to avoid (anaphylactic) reactions, but under this treatment most of the IgG molecules are split into an Fab and an Fc part and are no longer functionally active (unpublished observations). Therefore, after some time intramuscular gammaglobulin is often replaced by plasma (10 ml plasma per kg body weight per 2 weeks). Although the plasma derives from a single donor and thus usually has a smaller variety of antibodies, in practice this form of prevention is effective.

A very recent development in gammaglobulin therapy is the slow subcutaneous infusion of 16% gammaglobulin by means of a small portable pump by the patient himself at need. For instance, under increased exposure to respiratory pathogens or reduced resistance due to fatigue the dose or frequency can be increased.

REFERENCES

1. Römer PH: Weitere Versuche über Immunität gegen Tuberkulose durch Tuberkulose, zugleich ein Beitrag zur Phthisiogenese. In: Beiträge zur Klinik der Tuberkulose und spezifischen Tuberkulose-Forschung. Brauer L (ed), A. Stuber's Verlag, Würzburg, 1909, XIII, p. 1.
2. Landsteiner K, Chase MW: Experiments on transfer of cutaneous sensitivity to simple compounds. Proc Soc Exp Biol Med 48:688, 1942.
3. Chase MW: The cellular transfer of cutaneous hypersensitivity to tuberculin. Proc Soc Exp Biol Med 59:134, 1945.
4. Van Furth R: Some aspects of vaccination. Neth J Med 22:123, 1980.

DISCUSSION

Dr Gould: I was very interested to hear you referring to passive immunization with gammaglobulin. Dr van der Meer also mentioned it in connection with the immuno-suppressive patients. For a number of years, in association with my colleagues in the blood transfusion service in Edinburgh, we have been treating severely ill patients, some with respiratory infections, some with bacteriaemic shock when no organism has actually been isolated from the blood, and in some respiratory cases and cases with severe infection in burns, with massive doses of pooled human gammaglobulins, so that they receive intravenously as a rule, on occasion intraperitoneally, about 20 to 35 g of gammaglobulin per day. This does seem to turn the tide in some very severely ill patients, but, as you can well imagine, it is a very difficult subject to control properly, to get statistically significant results one way or the other. But we still believe this is something worth adding to the therapeutic armory, in selected cases.

You did mention that plasmaphoresis has been done in relation to well-defined infections, such as tetanus, to make available human specific immunoglobulin. For a time, we have been attempting to collect cases which have recovered from, for example, staphylococcal infection and these patients have their blood plasma frozen. Sera of a number of such patients are collected to give a small bank of what one would call hyper-immune human serum for administration to patients who, we think, are suffering from toxemia in addition to the bacterial infection. My suggestion was that this should be looked at in much greater detail, and particularly with Gram-negative infections.

It is obviously very important, but it seems rather difficult to measure, in serum from cases recovered from Gram-negative infections, the antitoxic antibody titer. I have tried to do it with samples of the human pooled globulin and I have tried to measure improvement in the killing power of the recipient's serum or plasma against the homologous organism, when this is available, but I cannot say that this has been in the slightest way successful. The only evidence we have is rather subjective on our part: we think that some people occasionally benefit from this therapy.

Dr Dijkman: Patients with chronic obstructive respiratory disease, because they are liable to get infections, often produce sputum in which pneumococci are present. These bacteria appear and disappear. The natural history of such patients includes the formation of antibodies and still recurrences of bronchial infections by pneumococci continue to occur. I wonder, what sense vaccination against pneumococci would make. Relapses of infectious periods in these patients, however, are not stopped, in spite of the presence of demonstrable antibodies. The question I am asking is: do you consider patients having chronic obstructive respiratory disease to be benefited by pneumococcal vaccine?

Dr Van Furth: No, at present I am not in favour of immunizing these patients. I think good clinical trials should be done to see what really happens when these patients are vaccinated.

Chairman: I wonder how British general practitioners are going on with the pneumococcal vaccine? Or are they not?

Prof. Grob: We generally advocate vaccinating our old and our sick with influenza vaccin. We have lists of risk patients. They seem to like it, but whether this is just because they feel it does them good and they like to see their doctor each November, I so not really know. But I am surprised at the enthusiasm among the patients.

The other groups that we tend to vaccinate are, of course, ourselves and our nurses. Perhaps I ought to ask whether you consider this another group that ought to be vaccinated? Certainly, when epidemics are around, you really need a few doctors around, too.

Chairman: My question was about pneumococcal vaccination.

Prof. Grob: This is a much harder one. There are divided opinions. I think we are suspicious of commercial motivation which seems to be pressuring us into the pneumococcal vaccines. We are guarded and not convinced is the answer.

Dr Gould: I think that in the United Kingdom, long before we would think seriously of pneumococcal vaccination, we would require to study once again the epidemiology of the biotypes. Currently, there is very little information available. I did mention earlier that we had recently a quite marked outbreak of pneumococcal infection, certainly with a lot of respiratory disease, but with spin-offs of a few cases of meningitis, arthritis, etc., which we do not normally see.

Chairman: Perhaps Dr Vanderpitte will add the Belgian view.

Dr Vanderpitte: The Belgian view chiefly is the absence of a view. This vaccine is peculiar in this sense, that the population, and also the medical population, does not feel the need for this vaccine. So, this need will probably be artificially created by some promotion campaign or other. I do not see myself how this promotion will function. Will the vaccine be advocated for elderly people, or will they limit the indications to selective or elective vaccination? But, in the last case, I feel that the market is too small to justify the enormous investment that is necessitated for such a tricky, and such a composite and complicated vaccine. So, I cannot foresee the commercial prospects of this vaccine in our country.

Dr Kerrebijn: How can we expect vaccination to be effective in immuno-suppressed patients?

Dr Van Furth: We do not know. We do know that influenza vaccin is effective in renal transplant patients. Bone marrow transplants are at risk for pneumococcal infection. However, the production of antibodies after vaccination can be induced only 1 or 3 months after bone marrow transplantation. These patients are very severely immuno-suppressed by irradiation and cytotoxic drugs before transplantation and it takes some time before enough B and T lymphocytes are formed to participate in an adequate immune response. In practice, one must start to vaccinate with DTP vaccin one month after transplantation and check whether antibodies are formed. If not, vaccination should be repeated. Only when the patient is again (partly) immunocompetent, influenza and pneumococcal vaccin should be given.

With regard to a future vaccin against cytomegalovirus, the situation is gloomy, because these infections occur often in the immuno-suppressed period after transplantation and then protective vaccination is still ineffective. One could consider to vaccinate the donor. Perhaps in the future we can vaccinate volunteers and after plasmaphoresis prepare hyperimmune gammaglobulin for passive administration. This might prevent the disease. The vaccination of other immunosuppressed patients, such as those with Hodgkin, non-Hodgkin lymphoma, acute leukaemia, etc. is not to predict. Vaccination of these patients with influenza and pneumococcal vaccin should be done if their physician expects that it will prevent illness. When splenectomy is done in these patients, I vaccinate them with pneumococcal vaccin.

Dr Sundberg: A large problem in otorhinolaryngology is the recurrent otitis media, due to pneumococci. In spite of adequate treatment, pneumococci seem to survive and persist in the epipharynx. In Scandinavia, now, they are studying the effect of pneumococcal vaccine to evaluate whether it is possible to attack the problem of recurrent otitis media in this way.

Dr Van der Waaij: Could we switch over to the Gram-negatives? More and more evidence is seen in the literature, that lipopolysaccharides, in particular doses from various Gram-negative micro-organisms, do cross-react with tissue antigens. Secondly, I would rather concentrate on an anti-lipid A-antibody for protection, rather than a more specific antibody which is specific to certain enterobacteriaceae with a given serotype, and therefore has a very narrow spectrum in its application.

Dr Van Furth: I know about your concern about the cross-reactivity with tissue antigens which might, as far as I understand you, induce auto-immune responses.

Chairman: There is one interesting observation of Lowell Young, who studied experimental Gram-negative infection in rabbits. He found that the pneumococcal vaccine gave protection against certain Klebsiella strains and other Gram-negative bacteria.

SUBJECT INDEX